THE EMPIRICAL TURN IN THE PHILOSOPHY OF TECHNOLOGY

RESEARCH IN PHILOSOPHY AND TECHNOLOGY

Series Editor: Carl Mitcham

Recent volumes

Volume 9: Ethics in Philosophy and Technology

Volume 10: Technology and Religion

Volume 11: Technology and Politics

Volume 12: Technology and the Environment

Volume 13: Technology and Feminism

Volume 14: Technology and Everyday Life

Volume 15: Social and Philosophical Constructions of Technology

Volume 16: Technology and Social Action

Volume 17: Technology, Ethics and Culture

Volume 18: Philosophies of the Environment and Technology

Volume 19: Metaphysics, Epistemology and Technology

RESEARCH IN PHILOSOPHY AND TECHNOLOGY VOLUME 20

THE EMPIRICAL TURN IN THE PHILOSOPHY OF TECHNOLOGY

GUEST EDITORS

PETER KROES
Department of Philosophy, Delft University of Technology, The Netherlands

ANTHONIE MEIJERS
Department of Philosophy, Delft University of Technology, The Netherlands

SERIES EDITOR

CARL MITCHAM
Colorado School of Mines, Colorado, USA

Emerald

United Kingdom – North America – Japan –
India – Malaysia – China

Emerald Group Publishing Limited
Howard House, Wagon Lane, Bingley BD16 1WA, UK

First edition 2000.

Reprints and permission service
Contact: books@emeraldinsight.com

British Library Cataloguing in Publication Data
A catalogue record for this book is available from the British Library

ISBN: 978-0-762-30755-5

Printed and bound in Great Britain by
CPI Antony Rowe, Chippenham and Eastbourne

EDITORS/EDITORIAL BOARD

Research in Philosophy and Technology is a refereed publication. Editorial correspondence should be addressed to Carl Mitcham, General Editor, Stratton Hall 301, Liberal Arts and International Studies, Colorado School of Mines, Golden, CO, 80401.

DEDICATED TO ANDRIES SARLEMIJN

CONTENTS

SERIES EDITOR'S NOTE

As general editor of the *Research in Philosophy and Technology series*, it is a special pleasure to turn RPT volume 20 over to guest editors Peter Kroes and Anthonie Meijers and to acknowledge their hard work over a couple of years. It was during the Spring semester of 1998 – which I spent in the Netherlands at Tilburg University, the Technical University of Twente, and Delft Technological University supported by the Dutch Research Organization – that I first became aware of the full extent of the Kroes-Meijers program for an "Empirical Turn in the Philosophy of Technology." Indeed, it was during this time, impressed with both their philosophical vision and practice, that I offered them the opportunity to guest edit an issue of *RPT.* I am pleased that they were able to take me up on this offer, and that they have brought this project to completion in the form of the present volume. I am confident that this programmatic collection will make a major contribution to advancing the field of the philosophy of technology.

Because of the significance and thematic unity of this volume, I am departing from the standard *RPT* policy of including a substantial number of reviews of relevant recent publications. To both our reviewers and readers, however, I note that reviews will continue to be included in future volumes. Indeed, in order to deal with a backlog of reviews, the next regular *RPT* volume will make a special commitment in this regard.

Let me also take this occasion to acknowledge and to thank our new publisher, Elsevier Science, and its staff – especially in the persons of Ann Marie Davenport (editorial) and Colin Parsons (production) – for their excellent support. I am pleased and honored that Elsevier, after purchasing JAI Press, which initiated the *RPT* series three decades ago, has chosen to continue to commit to this publication.

C.M.
August 2000

GUEST EDITORS' PREFACE

This collection of essays is written in the firm belief that the philosophy of technology should take technology much more seriously. Philosophers have to open the black box of technology and base their analyses on an inside view of the engineering practice and on empirically adequate descriptions of technology. Such an empirical turn will not simply provide illustrations of existing philosophical ideas and analyses, but will open up whole new areas of research. Methodological, epistemological, ontological, and ethical questions concerning technology and the technical sciences will eventually come into focus that so far have been addressed only marginally or not at all in philosophy of technology. The volume obviously cannot be but a small step in this direction, but it does provide a glimpse into what a future philosophy of technology might look like.

Earlier versions of most of the papers in this volume were presented at a workshop "The Empirical Turn in the Philosophy of Technology", in Delft (The Netherlands) in 1998. Participants from both the philosophy and engineering profession came together to discuss issues of mutual interest and future research. The workshop was organized by the Department of Philosophy of Delft University of Technology, and sponsored by the Netherlands Organization for Scientific Research (NWO). We greatly acknowledge this support.

The volume is dedicated to Andries Sarlemijn. He has been a staunch defender of a more empirical approach in the philosophy of technology. One of us – Peter Kroes – had the privilege of working with him for many years, while Andries was head of the department of Philosophy at the Eindhoven University of Technology (The Netherlands). He was invited to present a contribution at the workshop and although he was already seriously ill at the time, he nevertheless accepted with much enthusiasm. Unfortunately, his illness prevented him from finishing his paper; he died on the first of January 1998.

Delft, July 2000
Peter Kroes
Anthonie Meijers

INTRODUCTION: A DISCIPLINE IN SEARCH OF ITS IDENTITY

Peter Kroes and Anthonie Meijers

The final volume of *Philosophy and Technology* – the official publication of the Society for Philosophy and Technology – appeared in 1995 and bears the title *New Directions in the Philosophy of Technology* (Pitt, 1995). Five years earlier, volume 7 had appeared under the title *Broad and Narrow Interpretations of Philosophy of Technology* (Durbin, 1990). Volume 6, published in 1989, bears the title *Philosophy of Technology: Practical, Historical and Other Dimensions* (Durbin, 1989). These publications show that for more than a decade now, philosophers interested in the phenomenon of technology have been discussing the nature and scope of philosophy of technology. Apparently, this domain within philosophy has been suffering from a kind of identity crisis. From the 1960s to the late 1980s philosophy of technology was dominated mainly by metaphysical analyses of technology (under the influence of Heidegger), and by critical reflections on the consequences of science and technology for the individual and social form of life. During and after this period some philosophers of technology started searching for a reorientation of the field.

One of the most interesting developments in this connection is the emergence of proposals to shift attention from the consequences of technology to technology itself. This is most clearly illustrated in the book *Thinking through Technology: The Path between Engineering and Philosophy* by Carl Mitcham (1994). This book is one prolonged plea that philosophers of technology should take technology more seriously. Its aim is (p. 12) "to indicate the proper approach, basic conceptual distinctions, and fundamental problems within which a comprehensive philosophy of technology resides." And the proper way to do this is (p. 267) "in a way that does not exclude engineering discourse." In other

The Empirical Turn in the Philosophy of Technology, Volume 20, pages xvii–xxxv.
2000 by Elsevier Science Ltd.
ISBN: 0-7623-0755-2

words, philosophers of technology have to start a dialogue with engineers about their philosophical analyses of technology, and in order to do so they will have to learn the language of engineering. From a different perspective, more or less the same message is put forward in Joseph Pitt's book *Thinking about Technology: Foundations of the Philosophy of Technology* (Pitt, 1999). In his opinion, post-World War II philosophy of technology has been primarily "social criticism", i.e., "critical denunciations of the negative effects of technology on human values and on human life" (p. vii). Pitt argues that in the "order of knowing" there is a logical priority of epistemological issues concerning technology over social criticism. In his view, "Understanding what we know about technology, and understanding *how we know that what we know is reliable*, are the prerequisites to offering sound evaluation of the effects of technologies and technological innovations on our world and on our lives" (p. viii). Here, too, a turn toward technology may be discerned.

It is against this background that a workshop entitled *The Empirical Turn in the Philosophy of Technology* was organized by the editors of this volume at the Delft University of Technology, the Netherlands, 16–18 April 1998. The workshop was intended to contribute to the ongoing debate about the nature and task of the philosophy of technology. The guiding idea behind the workshop was that for an adequate understanding of the role of modern technology in society and its consequences on human and social life, it is not sufficient to look at technology only from the "outside". A philosophical reflection on these matters has to be informed by what goes on "inside" technology; it has to be based on an understanding of technological practice itself, of how technological artifacts are designed, developed, and produced by engineers. This means, for instance, that the philosophy of technology has also to encompass a philosophy of engineering design. In line with the ideas expressed by Mitcham and Pitt, the organizers of the workshop thought that it was time for philosophers of technology to open the black box of technology.

Opening the black box of technology sounds rather simple, but it is not. As Winner asks (1993, p. 363) "*Where* does one go to learn what one needs to know to write confidently about philosophy and technology?" Indeed, the list of possible locations is very long (p. 363):

> Where should a philosopher go to learn about technology? To a research and development laboratory? A farm? An electrical power plant? A communication centre? An airport? An arsenal? A construction site? Offices and any agency that funds research? A toxic waste dump? An automated theme park? A school where computers are being introduced? What does one's understanding of a specific location and specific varieties of technical apparatus, knowledge, and practice contribute to one's ability to talk in penetrating, reliable ways about modern technology in general?

With so many locations and different aspects, there is a real danger of oversimplifying the content of the black box called "technology".

Technology is not a monolithic entity. To begin with, the notion of technology itself has various meanings: primary among these are technology as artifacts, as a form of knowledge, and as a form of action (see Kroes, 1998 and Mitcham, 1994). But each of these forms of technology manifests itself in many complex varieties. For instance, technology as a form of knowledge has branched in our times into various engineering disciplines, such as architectural engineering, systems engineering, mechanical engineering, electrical engineering, physical engineering, software engineering, etc. So, the black box hides a richness of historically grown forms of technology[1] Philosophers interested in understanding the nature and role of modern technology cannot neglect this richness without running the risk of oversimplifying their object. Precisely in order to avoid that danger, the organizers of the workshop decided to bring together practicing engineers and philosophers interested in technology for a discussion of the possibilities of an *empirical turn* in the philosophy of technology. On the agenda of the workshop was particularly the relevance of empirical case studies (more generally, of empirical material) providing adequate descriptions of parts or aspects of modern technology for philosophical reflection on technology.

Our call for an empirical turn in the philosophy of technology was inspired by the idea that philosophical reflection on technology should not start from preconceived and simplistic images of technology; instead, philosophical reflection should be based on empirically adequate descriptions reflecting the richness and complexity of modern technology. But what does the expression "empirical turn in the philosophy of technology" mean? What are its implications for the philosophy of technology? What is the relevance of empirical material for philosophical reflection, and how can it or should it influence philosophical questioning of technology? Before we briefly explore a number of different interpretations of the expression "empirical turn in the philosophy of technology", a closer look at the classic dichotomy between philosophy and the empirical sciences is in order.

1. PHILOSOPHY AND EMPIRICAL SCIENCE

Traditionally, empirical material (matters of fact) is not considered to be the object of study of philosophy. Whereas the empirical sciences deal with synthetic truths that are grounded in matters of fact, philosophy is supposed to deal with matters of an analytic nature, i.e., matters concerning meaning independently of matters of fact. However, in his famous article "Two Dogmas of Empiricism", Quine (1963) convincingly argues that a sharp distinction between the domain of the analytic and of the synthetic is difficult to maintain. It is one of the two dogmas of empiricism that such a sharp and fundamental borderline exists. For Quine, the distinction between the analytic and the synthetic is a matter of degree, not of principle. The truth of statements depends on language (the analytic aspect) and on matters of fact (the synthetic aspect). But, he argues, in general it is not possible to analyze the truth of statements unambiguously into a linguistic and a factual component. Therefore, it is difficult to draw a boundary around analytic statements, that is, statements for which the factual component is nil. For Quine it is not the individual statement that has empirical significance, but the whole of our knowledge (pp. 42-43):

> The totality of our so-called knowledge or beliefs, from the most casual matters of geography and history to the profoundest laws of atomic physics or even of pure mathematics and logic, is a man-made fabric which impinges on experience only along the edges. Or, to change the figure, total science is like a field of force whose boundary conditions are experience. A conflict with experience at the periphery occasions readjustments in the interior of the field. Truth-values have to be redistributed over some of our statements. [. . .] But the total field is so underdetermined by its boundary conditions, experience, that there is much latitude of choice as to what statements to re-evaluate in the light of any single contrary experience.No particular experiences are linked with any particular statements in the interior of the field, except indirectly through considerations of equilibrium affecting the field as a whole.

Depending on the circumstances, any statement "can be held true come what may, if we make drastic enough adjustments elsewhere in the system" (p. 43). This means that any statement can be turned into an analytic statement. From this perspective, the classic dichotomy between philosophy and the empirical sciences is inappropriate. As Quine remarks, one of the consequences of giving up the dogma of the strict boundary between analytic and synthetic statements is "a blurring of the supposed boundary between speculative metaphysics and natural science" (p. 20). The difference between philosophy and the empirical sciences becomes one of degree. The idea that there exists a fundamental cleavage between philosophy and the empirical sciences has to be given up. With this observation in mind, we will now turn to various interpretations of the expression "the empirical turn in the philosophy of technology".

2. EMPIRICAL PHILOSOPHY OF TECHNOLOGY

The expression "empirical turn in the philosophy of technology" could easily be interpreted as an attempt to turn philosophy of technology as far as possible away from analytic matters and to focus it primarily on matters of fact. The result of the empirical turn would then be an "empirical philosophy of technology", in which one of the primary functions of empirical evidence would be to support (verify or falsify) particular statements. As such, it would be comparable to, for instance, the sociology of technology or the economics of technology. In other words, it would be part of technology studies, and within that domain it would be on a par with other empirical disciplines. Philosophy of technology would become an empirical enterprise, and lose its "philosophical" character.

We do not endorse this interpretation of the notion of an empirical turn in the philosophy of technology. On the contrary, philosophy of technology should keep its distinctive philosophical nature. Nevertheless, it should also base its analyses on empirical material, much more than has been done so far. Quine's image of human knowledge as "a man-made fabric which impinges on experience only along the edges" may help to clarify our notion of an empirical turn in the philosophy of technology. In terms of this image, the empirical turn may be interpreted as moving philosophical questions about technology away from the center of the fabric with all "its elaborate myths or fictions", toward the edges of the fabric, which "must be kept squared with experience" (p. 45). But it should not be moved right to the edges of this fabric, because that would turn it into an empirical science: There the primary concern is to keep the fabric in accordance with experience.

The philosophy of technology should concentrate more on the clarification of basic conceptual frameworks used in the engineering sciences and in the empirical sciences studying technology and less on abstract myths and fictions of which it is not clear how they relate to the real world of technology. This does not imply the transformation of the philosophy of technology into a "full-blooded empirical science", but empirical considerations become relevant for doing philosophy. Its primary aim is not empirical adequacy, but the clarification of the conceptual frameworks used in empirically adequate descriptions of technology. These descriptions regard technology as it is conceived in engineering practice, as well as technology as it functions in our everyday world.

3. A NON-NORMATIVE, DESCRIPTIVE PHILOSOPHY OF TECHNOLOGY

Another way to interpret the notion of an empirical turn in the philosophy of technology is that it should turn away from normative/evaluative matters toward empirical/descriptive ones. Such a turn could take place in one of two ways (see Diagram 1). On the one hand, it may take place with regard to the *approach* taken in the philosophy of technology. In this case, it implies that the field

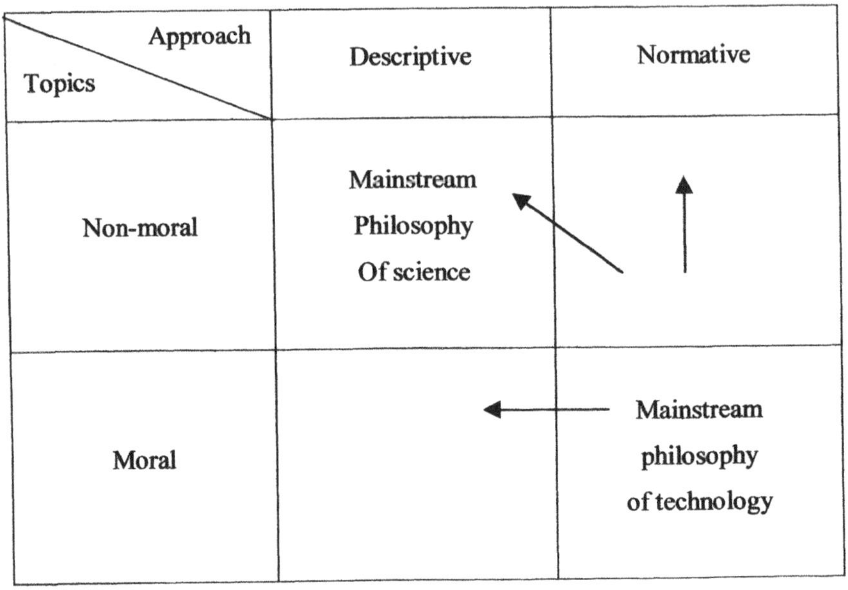

Diagram 1.

should take a more descriptive approach toward issues regarding technology instead of a normative/evaluative one (horizontal arrow). It is clear that much work done in this field since the Second World War has been dominated by a normative/evaluative stance toward technology, and very often a negative one (technology as some kind of threat to human existence). Normative/evaluative stances in the philosophy of technology usually are explicitly or implicitly of a moral nature. On the other hand, it may concern the *topics* studied. In that case, the turn implies that the philosophy of technology should shift its attention from moral issues concerning technology to issues of a non-moral, descriptive nature (e.g., epistemological, ontological, or methodological issues) (vertical arrow). The combined effect of these two turns is represented by the diagonal arrow in Diagram 1.

It is interesting to make a brief comparison with the philosophy of science. With regard to the approach taken and the topics studied, the philosophy of science is more or less the mirror image of the philosophy of technology. In the highest-ranking journals in this field (i.e., *Philosophy of Science, The British Journal for the Philosophy of Science*), virtually no attention is given to moral issues concerning science or the latter's effects on humankind and society[2]

Whenever the issue of normativity arises in the philosophy of science, it is usually with regard to the approach the philosophy of science should take with regard to science. Take as an example the Kuhn-Popper debate about the descriptive versus normative nature of philosophy of science. Here the issue is whether the philosophy of science should restrict itself to describing the way science is actually done, or whether it should also evaluate actual science from a methodological point of view. Note that in contrast to the normative/evaluative approaches in the philosophy of technology, normative approaches in the philosophy of science are not of a moral nature.

The issue about the role of normativity in the philosophy of technology is one of the controversial ingredients in the identity crisis mentioned above. For some, the approach of the philosophy of technology is inherently normative. For example, Paul Durbin (1978, pp. 67-68) argues that "what a particular philosophy of technology amounts to is a statement as to what one feels a good technological society *ought* [his emphasis] to be like". For others, the philosophy of technology is not inherently linked to a normative stance. For instance, most of the contributions to this volume take what could be called a descriptive, non-moral approach in the philosophy of technology.[3] Whether or not a normative or a descriptive approach is appropriate for the philosophy of technology strongly depends on what one considers to be the aim of this branch of philosophy. Such, of course, has been the focal point of much of the discussion about the nature and scope of the philosophy of technology in the last decades.

When we named our workshop *The Empirical Turn in the Philosophy of Technology*, we did not have this interpretation in mind. It was not our intention to argue that the philosophy of technology should follow the road taken by the philosophy of science and leave aside all matters of a normative (moral) nature.[4] In that case, the philosophy of technology would become vulnerable to the kind of criticism Langdon Winner raises against social constructivist theories of technology. These theories show a "lack of and, indeed, apparent disdain for anything resembling an evaluative stance or any particular moral or political principles that might help people judge the possibilities that technologies present" (1993, p. 371). Many of the pressing social issues regarding technology are of a moral nature, and it is almost self-evident that the philosophy of technology has also to address these.[5] But this does not imply that it has to be focused exclusively on such issues and should address them from a normative/evaluative stance.

Apart from moral questions, modern technology raises important methodological, epistemological, and metaphysical (ontological) questions. These questions are interesting in their own right. In order to understand better the nature of modern technology (e.g., in comparison to modern science, but also in comparison to earlier forms of technology) they have to be dealt with. These questions are not simply subordinate to or parasitic on normative/evaluative issues. On the contrary, they may be highly relevant for studying those issues. That is precisely what Pitt argues for when he claims "the logical priority of epistemological issues over social criticism in the 'order of knowing'" (1999, p. viii). In a similar vein, Mitcham has stated that more emphasis on fundamental theoretical issues dealing with technology may make it possible "to make a more profound contribution to ethical reflection than by immediate engagement with particular moral problems" (1994, p. 7). But whether epistemological or fundamental theoretical issues are involved, in both cases philosophical analysis has to be based on reliable, empirical descriptions of technology. Even if the ultimate aim of the philosophy of technology were to be to address moral issues with regard to modern technology, in many cases an understanding of modern technology itself would be necessary, and that would require an "empirical detour" in the philosophy of technology.

4. AN EMPIRICALLY INFORMED PHILOSOPHY OF TECHNOLOGY

Our call for an empirical turn in the philosophy of technology is thus not be understood as a wish to turn this branch of philosophy into an empirical science, nor as a plea to turn it away from normative/evaluative matters. As stated earlier, it is a call to base philosophical analysis concerning technology on reliable and empirically adequate descriptions of technology (and its effects). In our opinion, this is a *conditio sine qua non* for the philosophy of technology to be taken seriously in present-day discussions about technology. This does not mean that its primary focus should be on empirical problems, for that would turn it into an empirical science. Its focus should be on conceptual problems, more in particular, on the clarification of basic concepts and conceptual frameworks employed in empirically adequate descriptions of parts or aspects of technology. Such a turn would be more or less comparable to what has happened in the philosophy of science in the last decades. Inspired by the work of Kuhn, many studies in the philosophy of science have been carried out in recent years in order to develop a better understanding of the actual practice of scientists, particularly their experimental work in laboratories. Just as more emphasis on detailed case studies and on relevant work of historians of science has enriched the philosophy of science, so the philosophy of technology could profit from a similar change.

It should be stressed that, in our opinion, more is involved in an empirical turn than just *illustrating* or *supporting* existing philosophical ideas or analyses with more detailed empirical case studies. Nor is this turn simply a matter of *applying* philosophical ideas or results to technology. Such an interpretation of the empirical turn would imply that the kind of philosophical questions posed concerning technology would *grosso modo* stay the same; the sole aim of the empirical turn would then be to provide a firmer empirical grounding of the answers proposed in the actual practice of technology. We expect that the empirical turn will lead in various areas to problem shifts and to the emergence of new clusters of problems. Our idea is that looking at technology through the magnifying glass of detailed empirical case studies will reveal new topics and new conceptual frameworks that will give rise to their own, specific kinds of philosophical issues. To illustrate this point, we will briefly sketch below the kind of topics and conceptual problems that will be studied in a research program called "The Dual Nature of Technical Artifacts".[6] This program is part of an international research program called *Philosophical Foundations of Modern Technology*.[7] The aim of this international program is to analyze and develop basic concepts and categories necessary for understanding modern technology (first of all, the central notions of technical function and engineering design) and to integrate them into a coherent conceptual framework.

Our program on "The Dual Nature of Technical Artifacts" starts from the observation that technical artifacts have a twofold nature: They are (i) designed *physical structures* which realize (ii) intentionality-bearing *functions*. This duality is related to the fact that we employ two basic conceptualizations of the world in our thinking, speaking, and acting. On the one hand, we see the world as consisting of physical objects interacting through causal connections. On the other, we see it as consisting of agents (primarily human beings), who intentionally represent the world and act in it. *Both conceptualizations are necessary for characterizing technical artifacts.*

Technical artifacts are, *prima facie*, always physical objects, but they are also objects which have a certain function. Looked upon as merely physical objects, they fit into the physical or material conception of the world. Looked upon as functional objects, however, they do not. The concept of function never appears in physical descriptions of the world; it rather belongs to the intentional conceptualization. This is shown, for one thing, by the fact that attributions of function lend themselves to normative judgments – artifacts can perform their function well or badly – and normative statements make sense only within the intentional conceptualization. Technical artifacts thus have a dual nature: They cannot be described exhaustively within the physical conceptualization, since this has no place for their functional features, nor can they be described exhaustively within the intentional conceptualization, since their functionality must be realized in a physical structure that is adequate to it.

Exactly how the intentional and the physical conceptualizations are related is a fundamental problem. If, on the one hand, functions are seen primarily as "added to" the physical substrate, or as realized in physical objects, then the question remains how these functions are related to the mental states of human individuals – which, after all, form the core of the intentional conceptualization. If, on the other, functions are seen primarily as patterns of mental states, and exist, so to speak, in the heads of the designers and users of artifacts only, then it becomes somewhat mysterious how a function relates to the physical substrate in a particular artifact. Given this state of affairs, the general aim of the research program is to develop a coherent conceptual framework for describing technical artifacts, that takes due account of their dual nature.

In order to realize this general aim, various topics related to the dual nature of technical artifacts have to be studied in conceptual and empirical detail. The following list gives a first impression. It will be necessary to give an in-depth conceptual analysis of the notion of technical function as it is used in engineering practice; to explore the possibilities of a taxonomy of technical functions; to compare this account with the accounts of function given in other areas, especially biology and the social sciences; to develop an account of the relation between technical function and physical structure, particularly the sense in which the physical structure constrains the technical function; to develop an account of the process of *designing* technical artifacts, and of the way the structural description of the artifact as physical object and the functional description of the object as intentionally formed artifact are combined during this process (How do engineers succeed in bridging this gap?); to explore the possibilities of a rational reconstruction of designing technical artifacts, and particularly of the evaluation of design; and to examine how technical explanations in design are related to the various types of explanation that are traditionally distinguished in the philosophy of science.

Furthermore, it will be necessary to develop an account of the intentional aspects of technical functions; to analyze the conceptual relation between the intentions, beliefs, and desires of *designers* and *users*, on the one hand, and technical functions, on the other; to investigate how designers communicate their intentional states to potential users; to analyze the normative judgments that apply to the functioning of artifacts, as well as the origin of this normativity; to explore the kind of social constraints involved in the design and use of technical artifacts; and to clarify in what sense(s) technical artifacts (telephones, cars) are different from such social artifacts as laws, universities, and money.

These issues also raise more general philosophical questions, in particular epistemological and ontological ones. In what sense is knowledge of functions different from knowledge of physical structures? Is it necessary to develop a separate epistemology of function attributions? Do technical artifacts have functional part-whole relations that are different from structural ones? Is the artifactuality and functionality of artifacts a relational property, and if so, what are its relata (individuals, social groups, other artifacts)? What are the identity criteria for technical artifacts, and what types of ontological commitments are involved in describing and using artifacts?[8]

These topics and questions show that a focus on technical artifacts, more in particular on their dual nature, introduces a host of interesting philosophical problems – problems that so far have been addressed only marginally or not at all in the philosophy of technology. It is obvious that a proper philosophical analysis of many of these topics can only be performed on the basis of empirically adequate descriptions of, for instance, various aspects of engineering design. More generally, a shift of focus toward technology within the philosophy of technology will make an empirical turn within the field inevitable.

The idea of an empirical turn in the philosophy of technology was explored in more detail at the Delft workshop. Speakers from both the philosophy and engineering profession came together to discuss issues of mutual interest and future research. The interdisciplinary nature of the meeting made it clear again that philosophers and engineers often live in different conceptual worlds and that success in interdisciplinary communication can not be taken for granted. At the same time cooperation and communication across disciplines is a sine qua non for a philosophy of technology that wants to be empirically informed.

The papers presented at the workshop addressed a great variety of issues. Some aimed at clarifying particular ontological, epistemological or ethical issues related to technology or to specific technological cases, while others discussed the empirical turn in more general terms. They all attempted to contribute to the idea of an empirical turn and many explored new ways of taking into account empirical data in their philosophical analyses. Revised versions of these papers have been collected in the present volume. It therefore reflects the very richness and variety of the contributions made during that stimulating meeting.

5. THE PAPERS IN THIS VOLUME

The volume starts with three chapters that address in more general terms the issue of an empirical turn in the philosophy of technology. In his contribution, "There's No Turn Like the Empirical Turn", Arie Rip maintains that an empirical turn is already visible in the philosophy of technology, but that it could use a larger input from reflective technology studies. This is particularly true when the philosophy of technology takes into account not only design and engineering practices, but also the wider world of which these practices are a part. In addition to traditional sociological and political science analysis and understanding of the wider world, technology studies focus specifically on the sociotechnical aspects. The two main issues discussed by Rip are (i) the important role of projections and promises which are an integral part of the introduction of technological novelty, and (ii) the embedding of technological innovations in sociotechnical regimes and a sociotechnical landscape. The discussion of these issues shows that Rip proposes not only an empirical input in the philosophy of technology, but also a research agenda. He illustrates this with a brief discussion of methodology and ontology.

Peter Kroes argues that the philosophy of technology is in bad shape in his paper, "Engineering Design and the Empirical Turn in the Philosophy of Technology". The field lacks coherence, and there is no consensus about its aim, topics to be addressed, or methodology. Characteristic of much work done is the externalistic approach to technology: The focus is on the social consequences of the use of technology. Technology itself remains a black box. He argues that the philosophy of technology should take technology seriously and address philosophical issues concerning design and engineering. He claims, moreover, that a fruitful philosophical analysis of problems related to technological design and engineering is only possible on the basis of detailed, empirical case studies. This is illustrated with an investigation of the relation between structure and function in the case of early steam engines.

In his contribution, "To What Extent Should a Critical Philosophy of Technology be Constructivist?", Bram Bos addresses the relevance of constructivist studies on technological change for an empirical turn in the philosophy of technology. Although these two fields of inquiry have worked in separate domains for decades, without many interactions between them, a confrontation would be fruitful, at least for a philosophy of technology that aspires to reach a critical position toward actual technological artifacts and systems. A constructivist ontological perspective of technology might be a good starting point for addressing the intimate linkages between material artifacts and social relations, and thereby highlight political and normative features of technology that might otherwise remain hidden. However, the constructivist methodology for doing empirical research is primarily oriented toward processes of change, and is therefore a useful but certainly not a sufficient resource for identifying the political and normative roles that concrete instances of technology might play. This theme is discussed, on the one hand, by way of analysis of the positions of Langdon Winner and Wiebe Bijker on the relevance of constructivism for "practical" issues and, on the other by reference to the attempt of Andrew Feenberg to develop a critical version of constructivism.

The next two chapters address issues of an *ontological* nature. Taking as a starting point his empirical, ethnographic studies of the design process, Louis Bucciarelli investigates the ontology of technical artifacts in his contribution, "Object and Social Artifact in Engineering Design". In *Designing Engineers* (1994), he argued that the design process is a social process, and that its final product – the artifact – is a social construct. Each individual involved in a design process works within her own *object world*, and there is no over-arching object world for settling disputes among individuals living in different object worlds. Bucciarelli explores in his paper the idea that technical artifacts do not exist as objects *per se*, and encourages the reader to come down from the mountain and see the social entanglements in the design and use of technical artifacts. For instance, he argues that the properties of a new material (AL-MMC) are social constructs, in the sense that they are only usefully defined within particular social contexts. Similarly, there is no such thing as the photovoltaic module: such a module has many faces, depending on the object world of the individual involved. Finally, Bucciarelli puts forward the conjecture that during the design process, the artifact to be designed exists in a collective sense. The community of designers has it in mind and keeps it energized in their collective memory during the design process. As such it should be regarded as a social artifact.

In his essay, "The Relational Ontology of Technical Artifacts", Anthonie Meijers distinguishes three *kinds* of properties of technical artifacts: intrinsic, context-dependent, and relational properties. His analysis focuses on the nature of technical functions, being a main property of artifacts. He argues that theories of function in the philosophy of biology are of limited use for the analysis and understanding of technical functions. There is one exception though. Many philosophers of biology make a distinction between proper function and system function, and this distinction is useful in the artifactual domain. His analysis shows that technical functions are relational properties. A function relates the artifact's physical structure to a practice of design and use. Since relational properties are irreducible, Meijers concludes that artifacts are relational entities. This implies that there is no such thing as an artifact in itself. It also follows that artifacts do not have clear boundaries and that there is often no unique answer to the question: How many parts does an artifact have? His analysis supports the main idea defended in this introduction, viz. that the practice of design and use should be taken much more seriously in the philosophy of technology.

The next three chapters deal with *epistemological* issues. Davis Baird argues in his paper, "The Thing-y-ness of Things: Materiality and Spectrochemical Instrumentation, 1937–1955", that one of the lessons of an empirical turn should be to focus more on the materiality of technological objects. In his opinion, historians and philosophers of technology have focused too exclusively on *ideas*, neglecting the fact that technological artifacts are primarily *material things*. Dealing with material things, however, is different from dealing with ideas: Every real screw is an imperfect realization of the idea of a perfect screw, and its imperfection has to be taken into account and compensated for in engineering practice. Starting from the thing-y-ness of things, Baird draws a number of lessons, which he illustrates with the example of the development of spectro-chemical instrumentation.

Clive Dym and Philip Brey's paper, "Languages for Engineering Design: Empirical Constructs for Representing Objects and Articulating Processes", deals with the nature of design knowledge and the cognitive structure of the design process. They start with an analysis of the nature and the role of external representational media or languages, as they call them, in design activity. By external representational media they mean representational forms that can be found outside the human mind. Their intention is to arrive at a typology of these representational forms and to assess their role in design. They also discuss the possibility of automating design activity through the design and manufacture of expert systems for product design. They take a detailed look at one of the most advanced systems of this sort, the so-called PRIDE system, and with the help of this example they explore the possibilities and limitations of automating design through the use of expert systems.

In his provocative paper, "Design Mistakes: The Case of the Hubble Space Telescope", Joseph argues that philosophy of technology has to begin with facts, not ideology (the social critics of technology) or metaphysics, examining what these facts tell us about how well our theories represent real world technology. Of course, facts do not speak for themselves and are always theory-laden. But this just emphasizes the importance of developing criteria for evaluating empirical claims about technology. Such criteria must take into account the way that our knowledge of technology is constrained by the methods, assumptions, and values we or others bring to our investigations. Once we are able to assess the empirical adequacy of our theories, we can unmask their ideological elements.

Pitt has a particular ideology in mind: the myth of the engineer, according to which engineers are paradigm cases of rational and project-oriented problem solvers. He discusses, then, three of theoretical models of technology (Vincenti's, Bucciarelli's, and his own) and confronts them with the facts of an interesting case in which the design went wrong, namely the case of the Hubble Space Telescope. Focusing on failure rather than success, Pitt shows that the models developed by Vincenti and Bucciarelli are only partially helpful. Their models are unable to explain or understand *failure* in design, and they seem to imply that in case of failed projects the participants are simply not very good engineers. In contrast to the myth of the engineer as idealized rational problem solver, then, Pitt proposes another point of departure in the analysis of technology, viz. the decision making process. Technology is humanity at work. We have to look at "who made what decision and why". The analysis of the Hubble case shows, in Pitt's view, that his own model is better suited to deal with the complexities of engineering design. The moral for an empirical turn in the philosophy of technology is that we need to focus much more on people. We need a better understanding of why people do what they do in their engineering practices.

The final part of the volume deals with *ethics.* It starts with a contribution by Carl Mitcham and René von Schomberg, titled "The Ethics of Engineers From Role Responsibility to Public Co-Responsibility". In their paper they consider the very emergence of an engineering ethics associated with philosophy to be a sign of an empirical turn in the philosophy of technology. Surveying the long history of ethics based on role responsibility, as a mediation between ethical systems and social structures, Mitcham and Von Schomberg argue that this kind of ethics was progressively abandoned by modern philosophy in favor of an ethics based on principle responsibility. Empirical evidence for the inadequacy of this shift can be found in the grassroots and largely non-philosophical efforts to revive role responsibility in engineering ethics. But in their opinion such a revival is insufficient, since there is a need for the development of something beyond role responsibility in modern, advanced technoscientific societies. They propose a new form of responsibility ethics, which they call collective co-responsibility ethics, and they indicate some of its central features. It is a new attempt to mediate between ethical systems and social structures.

The next four chapters are a dialogue between Henk Zandvoort and Ad Vlot on the relevance of legal measures for avoiding negative consequences of technology. In his paper, "Codes of Conduct, the Law, and Technological Design and Development", Zandvoort sets the stage for the discussion by addressing the issue of the socially responsible behavior of engineers, and of the corporations that employ engineers, especially with regard to sustainability and technological risks. He argues that neither codes of engineering ethics nor codes of corporate conduct are effective means for fostering socially responsible behavior among engineers, respectively in corporations. He points out that the duty of engineers, as laid down in most codes of engineering ethics, to inform the public on possible negative consequences of their work as a rule conflicts with labor contract law. Codes of corporate conduct turn out to be ineffective, because they are phrased in vague terminology and because of lack of systematic and independent control. According to Zandvoort, major legal changes are necessary. Conflicts between codes of engineering ethics and legal systems should be removed and, in order to encourage individual engineers and corporations to behave in a way that is consistent with sustainability and the responsible management of technological risk, legal systems should incorporate a form of what he calls strict liability. Finally, he puts forward the suggestion that stricter forms of liability may result in more scientific research and technological design and development rather than less, as is often assumed.

Ad Vlot confronts these claims with his personal struggle to be an ethically responsible human being as well as a professional aerospace engineer. He is involved in the process of developing a new material for the hull of airplanes. In his response, "Toward a Juridical Turn for the Ethics of Technology? An Aerospace Case", he argues that Zandvoort's advocacy of a legal turn will not contribute to solving ethical problems. Strict liability will not stimulate sustainability and protection of the environment, because it is not possible to determine who is liable for the greenhouse effect or the holes in the ozone layer. Safety will also not be promoted by strict liability; in his opinion cooperation between the triangle of manufacturer, airlines, and government agencies is crucial for safety. Finally, he argues that secrecy arrangements are vital in his daily work, and that the problem of whistle blowing is not the insufficient legal protection of the employee involved, but the gathering of enough reasonable evidence to blow the whistle. All in all, Vlot concludes that legal action will not solve the ethical problems he encounters in his daily work as an aerospace engineer.

In his reply, "Self Determination, Strict Liability, and Ethical Problems in Engineering", Zandvoort discusses why the legal changes he proposes (i.e., strict liability and the right of everyone to be informed on risks caused by others) are relevant for diminishing the ethical problems identified by Vlot. Furthermore, he argues that these legal changes are feasible. He does so by presenting an overview of developments in recent risk and environmental liability law, which shows that strict forms of liability and related concepts are already being practiced at some places in actual positive law. Zandvoort stresses that his proposals require far-reaching changes in several areas of contemporary legal systems, such as liability law in general, corporate law, labor contract law, etc. But there is no reason to assume that these changes will pose insurmountable difficulties. In his opinion, these changes are necessary in order to create more consensus about (and thus fewer ethical problems concerning) technological activities, among both the actors involved and those subjected to the consequences.

In his final comment, "Toward a Juridical Turn for the Ethics of Technology? Final Remarks", Vlot once more illustrates the ethically complex character of the day-to-day work of an engineer by discussing two examples of ethical problems which emerged after he had written his first comment on Zandvoort's paper. He admits that a legal turn may contribute in some cases to avoiding technological risks, but he is not convinced that it is a panacea for dealing with all risks induced by modern technology. Vlot is not optimistic about the possibility of creating a tremendous bookkeeping system to deal with risks in our technological society. In his opinion, much more emphasis should be given to social and cultural aspects of technical artifacts in engineering education. Moreover, the modern engineering design method is seriously flawed, because it lacks a direct link with the manufacture and use of technical artifacts.

CONCLUDING REMARKS

The chapters in this volume explore the various ways in which empirical data can be used in ontological, epistemological, ethical or more general discussions in the philosophy of technology. Together they show that the methodology of using empirical data in a philosophical analysis needs further reflection, in particular the criteria for selecting good case studies. All the papers in this volume attempt to contribute to an empirical turn in the philosophy of technology.

We have distinguished three different interpretations of an empirical turn and argued for a philosophy of technology that takes technology much more seriously. The philosophy of technology should be based on empirically adequate descriptions of technology and the engineering practices. The empirical turn in the philosophy of technology should not be confused with turning this branch of philosophy into an empirical science, for example as part of science and technology studies. Philosophy of technology differs from the empirical sciences in that it focuses on the clarification of basic conceptual frameworks used for describing and understanding technology, among other things. Furthermore, the empirical turn does not imply abandoning evaluative and normative questions. On the contrary, it is hoped for that a better understanding of technology resulting from an empirical turn will contribute to better normative analyses and evaluations. Finally, we have also claimed that the empirical turn will not just provide illustrations for existing philosophical ideas and theories, but will open up whole new areas of research.

An empirical turn is not easy to accomplish. If we want to base philosophical reflection on empirically adequate descriptions, a dialogue between philosophers and engineers will be necessary. However, communication across disciplines is notoriously difficult, especially in this case. The Delft workshop clearly showed that. The conceptual schemes, languages, intellectual backgrounds and passions in philosophy and engineering are indeed very different. Bridging this gap is the real challenge for philosophers of technology. The present volume is obviously but a small step in the direction of an empirical turn. It gives us a glimpse of what a future philosophy of technology might look like.

NOTES

1. Given this state of affairs, it is not surprising that in his *Introduction* to the volume *New Directions in the Philosophy of Technology*, Pitt remarks that (1995, p. viii), "Not everything new and exciting in the world of philosophy of technology is represented here. We are missing examples of philosophical dealings with engineering, for example."

2. C. West Churchman (1994), Editor-in-Chief 1948–1957 of *Philosophy of Science,* makes the following observation about the content of this journal in a guest editorial (p. 132): "The journal seems to spend most of its pages on the puzzles and imperfections of scientific theories, especially those arising in that most confusing of all the disciplines, physics. Whether human beings should study physical nature, whether it is dangerous or ethical to do so, how such studies relate to other human interests and activities; in fact, all the issues complicating the study of physics are never discussed in the journal."

3. Of course, this does not imply that all authors would defend a non-normative approach to the philosophy of technology in general.

4. The inclusion of contributions dealing with moral issues in this volume clearly shows that this is not the case.

5. Note that it may do so in a descriptive or normative manner, in line with the distinction between descriptive and normative ethics.

6. This four-year research program will be carried out at the department of Philosophy of the University of Technology of Delft, the Netherlands, starting in 2000. It consists of two post-doc projects and two Ph.D. projects, and is mainly funded by the Dutch Organization for Scientific Research (NWO). An extensive description of this research program is available on the website www.fil.tbm.tudelft.nl/. The description of the research program presented here is based on the program description which was jointly written by Peter Kroes, Anthonie Meijers, Maarten Franssen, Pieter Vermaas, and Wybo Houkes.

7. Apart from researchers from the University of Technology of Delft in the Netherlands, researchers from the USA (MIT, Virginia Tech, Georgia Tech, University of Buffalo and Caltech) will participate in this program.

8. In the research program referred to these topics will be further elaborated in four projects. Project 1 (postdoc) will investigate the various conceptual and ontological relations between technical function and physical structure. Project 2 (Ph.D.) aims at a rational reconstruction of technical design as a process of establishing basic relations between technical function, physical structure, and intended use. Project 3 (postdoc) will focus on the intentionality of technical functions and its consequences for the epistemology of function attribution. Finally, project 4 (Ph.D.) will analyze the necessary social conditions of technical functions, in order to further clarify their intentional nature.

REFERENCES

Churchman, C. W. (1994). Guest editorial: What is Philosophy of Science? *Philosophy of Science,* 61, 132–141.

Durbin, P. T. (1978). Toward a Social Philosophy of Technology. *Research in Philosophy and Technology* (pp. 67–97). Greenwich, CT: JAI Press.

Durbin P. T. (Ed.) (1989). Philosophy of Technology: Practical, Historical and Other Dimensions. *Philosophy and Technology*, 6. Dordrecht: Kluwer Academic.

Durbin, P. T. (Ed.) (1990). Broad and Narrow Interpretations of Philosophy of Technology. *Philosophy and Technology*, 7. Dordrecht: Kluwer Academic.

Kroes, P. A. (1998). Philosophy of Technology. In: E. Graig, (Ed.), *Encyclopedia of Philosophy*, *9* (pp. 284–288). London: Routledge.

Mitcham, C. (1994). *Thinking through Technology: The Path between Engineering and Philosophy*. Chicago: University of Chicago Press.

Pitt, J. C. (Ed.) (1995). New Directions in the Philosophy of Technology. *Philosophy and Technology*, 11. Dordrecht: Kluwer Academic.

Pitt, J. C. (1999). *Thinking about Technology: Foundations of the Philosophy of Technology*. New York: Seven Bridges Press.

Quine, W. van O. (1963) [1951]. Two Dogmas of Empiricism. *From a Logical Point of View: Logico-Philosophical Essays* (pp. 20–46). New York: Harper Torchbooks, ch. 2.

Winner, L. (1993). Upon Opening the Black Box and Finding It Empty: Social Constructivism and the Philosophy of Technology. *Science, Technology and Human Values, 18*(3), 362–378.

PART I

GENERAL CONSIDERATIONS OF THE EMPIRICAL TURN

THERE'S NO TURN LIKE THE EMPIRICAL TURN

Arie Rip

Philosophers of technology now turn to the phenomena in order to learn from them – always, and unavoidably, selectively. This is important and, as the essays in this book testify, productive. Learning from empirical work is not limited to turned-around philosophers, however. The actual work done in the "empirical turn" and the resulting insights could also result from sociologists and other empirical scholars "turning" to philosophy. One could call this "reflexive" sociology, and consider it to be a matter of inclination, of individual sociologists with a reflective bent (my favorite example here is Brown, 1989). But I would argue that it can and should be an integral part of sociology. When attempting to synthesize a large body of empirical and analytical studies, for example, one is forced to consider the variety in the object studied and inquire into its nature. When we did our review study of technological change (Rip & Kemp, 1998), we had to start with an analysis of various conceptions of technology and offer a considered view about them to support the approach we took in the review. That section of our paper could be part of an article in the philosophy of technology without anybody suspecting a sociologist and an economist to be behind it.

There is an added value of introducing sociology (reflexive or otherwise) in the philosophy of technology, in that it can highlight complexities and increase the scope of the analysis. This is how I will position my contribution, because in most chapters in the present book, the empirical turn is a turn to practices, in particular design and engineering practices. While this is OK as a starting point for philosophy of technology, it also introduces limitations. For one thing, the concentric view of technology – which is part of the culture of engineers

The Empirical Turn in the Philosophy of Technology, Volume 20, pages 3–17.

ISBN: 0-7623-0755-2

– may well be reproduced uncritically in this way, and a message may be conveyed that to understand technology it is enough to understand engineering.

Accepting the starting point, I shall add to it to overcome such limitations. A participant at the Delft Workshop, Pieter Adriaans, director of Syllogic, noted: "Society is programmed by the artifacts it creates itself." Clearly, there is much more to technology than the engineers' doing, even if they may carry a particular responsibility. The wider world must be introduced in the "empirical turn". And this can usefully be done by drawing on the field of technology studies, which has come into its own over the last two decades and has produced important (even if still somewhat fragmented) insights.

I will start inside the engineering world, then consider the different levels at which technology is embedded in society, and return to the engineering world by inquiring into its boundaries and partial rationalities. In the concluding section I draw some explicit implications for philosophy of technology.

ENGINEERS AND THEIR ARTIFACTS – AND HEURISTICS AND PROMISES

Inside the engineering world, the main actors are the engineers themselves, and their business is to design and build artifacts, whether components or complete systems, and to assure their continued working. It is on this base line that the self-image of engineers rests, and how they define their tasks and responsibilities.

Technology then appears as *configurations that work* – which is actually useful as a general definition, if one takes a broad interpretation of "configuration" (Rip & Kemp, 1998). These configurations are often realized sequentially, starting with some hardware, then adding the software of operating it (skills, logistics), the orgware of getting its production and use right, and increasingly also the "socioware" necessary for its embedding in concrete societal contexts. This concentric view allows engineers to define and defend their work style. It also implies that there is a boundary where their mandate and responsibility ends and others have to take over. The quest for such a boundary (and its eventual location) will then create a demarcation of technology (the engineer's concern) and its use and embedment in society (society's concern). The division of labor (and accountability) can be productive, but also allow gerrymandering – as is brought out nicely in Ravetz's (1975) aphorism for science: "Science takes credit for penicillin, while society is blamed for the bomb."

The boundary between technology and context is fuzzy, already in seemingly concrete cases as underground transport: where does the metro end and the city begin – and thus the responsibilities of the various actors (Achterhuis et al.,

1995)? And even more importantly: the boundary is constructed dynamically, between different and heterogeneous actors, and never conclusively. This is particularly clear for novel technologies such as biotechnology and the fusion of information and communication technologies, which are carried by promises (of the developers) and high expectations (of other actors). As the example of the metro indicates, it will also be important in infrastructural technologies. Again, promises and expectations will play their role, but now more focused on concrete functionalities.

The importance of promises and expectations is not limited to boundary work. It implies a general point about technology and engineering: on top of the base-line of actors and artifacts, "agenda" should be added to make our understanding of the dynamics of technological developments complete. We proposed a "Triple-A Triangle" – actors, artifacts, and agendas – to broaden existing technology studies (Van Lente & Rip, 1992; Van Lente, 1993), but such a triangle is equally important for a philosophy of technology that wants to profit from the empirical turn. Let me elaborate this, starting from technological design and ending up with general philosophical considerations.

Design necessarily starts with projections of what could be: an idea, a first sketch, and a story about how and why it would work if developed and realized. Agenda dynamics are immediately visible here in two ways. Firstly, expectations are formulated about what could be and what it would look like, and this sets out an agenda for further work. Such expectations can take the form of strong convictions and/or be supported by regularities in earlier developments. Moore's Law stipulating the doubling of processing speed and memory capacity in integrated circuits every one-and-a-half years is a striking example, which defines challenges for new technological development. As a news report in *Science* (1996) phrases it: "Researchers around the globe are working furiously to extend the life of Moore's Law by coming up with alternative chip-patterning techniques for use when current lithographic tools hit the wall." (And it is exactly because of such efforts, driven by innovation competition, that advances in chip technology remain predictable. As soon as firms decide to adopt another strategy, and go for alternatives, Moore's Law would lose its hold, and thus its validity.)

Secondly, the way to go about design and development is shaped by earlier experience, the heuristics of the dominant design paradigm (or paradigms) and other rules and standards of what has been called the technological regime. Nelson and Winter (1977, p. 57) refer to

> technicians' beliefs about what is feasible and worth attempting. For example, the advent of the DC-3 aircraft in the 1930s defined a particular technological regime: metal skin, low wing, piston powered planes. Engineers had some strong notions regarding the potential of

this regime. For more than two decades innovation in aircraft design essentially involved better exploitation of this potential; improving the engines, enlarging the planes, making them more efficient.

There are further agenda dynamics, though, which are often backgrounded – by engineers when they talk about their ongoing work, and by philosophers of engineering when they analyze what is going on. Promises are made to potential sponsors, and to the wider world at large, about what the specific design, or a new technological option, or a new project, will achieve. Such promises range from the mundane to the grandiose, but what they have in common is that they intend to mobilize resources for the project: funds and other material resources, as well as symbolic resources which create legitimation to proceed. One can see this happen in the present hype around internet companies and e-commerce and, in parallel, with mobile telephony. "The penetration of mobile phones will be more than 100% by 2003 – this is completely certain, there is no doubt about it." I have heard spokespersons from the telecom world emphasize this, and immediately draw out consequences for their, and others', strategies (Morley, 2000).

Promises made in public bind the speaker to some extent, and even more so when resources are acquired on the basis of the promises. As Van Lente (1993) has shown, promises mobilize resources which enable engineers to work on their project, but create requirements on what they must deliver, and thus introduce constraints on their work agenda. While a protected space is created where their work can continue, they have to work within the limits and limitations set by that space.

Thus, when a design is worked out and realized, it is not the linear unfolding of a technical possibility or the filling in of a functionality. It is an innovation journey, with its twists and turns, through which agenda dynamics materialize. And this is more than a self-fulfilling prophecy: there are continual feedbacks from experience, and these are assessed in relation to the effort actors are prepared to invest in further attempts. As Van Lente (1993) has shown for the case of a new polymer to be used as isolation in high-voltage cables, the assessments may shift over time, and lead to a (sudden) redefinition of the project and its prospects.

The innovation journey continues to pilots and demonstrations, market introduction, and niche building. The overall perspective about the role of heuristics (in existing regimes) and promises (to create openings) will again be applicable. Elsewhere, we have described a way to map the innovation journey and the way the halo of expectations gradually materializes (Rip & Schot, 1999). Figure 1 offers a (slightly adapted) version of the visualization used in that paper. For the details of the Figure, I refer to Rip and Schot (1999).

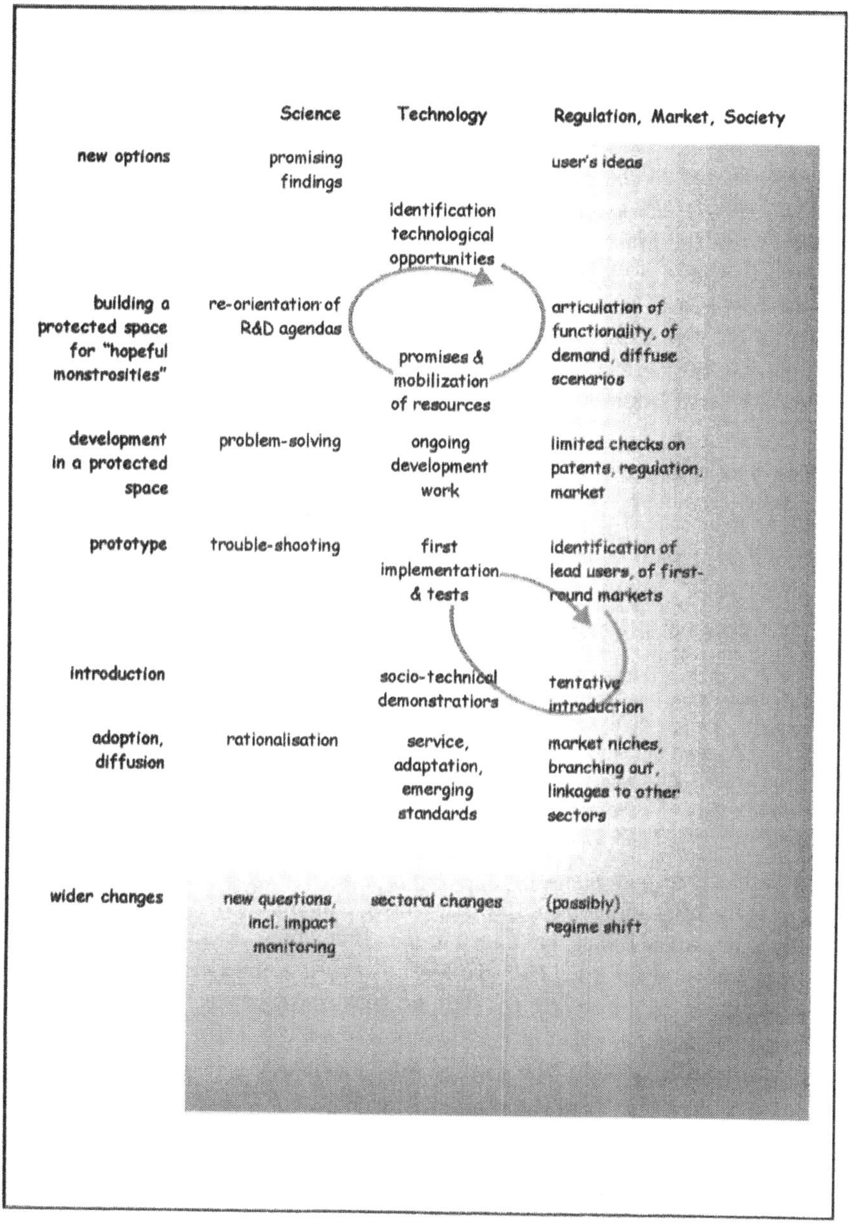
Science
Technology
Regulation, Market, Society
new options
promising findings
user's ideas
identification technological opportunities
building a protected space for "hopeful monstrosities"
re-orientation of R&D agendas
articulation of functionality, of demand, diffuse scenarios
promises & mobilization of resources
development in a protected space
problem-solving
ongoing development work
limited checks on patents, regulation, market
prototype
trouble-shooting
first implementation & tests
identification of lead users, of first-round markets
introduction
socio-technical demonstrators
tentative introduction
adoption, diffusion
rationalisation
service, adaptation, emerging standards
market niches, branching out, linkages to other sectors
wider changes
new questions, incl. impact monitoring
sectoral changes
(possibly) regime shift

Figure 1.

The next step in this reflexive sociology of innovation is to recognize the ontological elements involved. Engineers add to the furniture of the world, and will thus shift its ontology, in the small and in the large. Over time, major shifts may occur, as is evident if one traces the history of electricity from the late 19th century onward, and is the present, and shared, expectation about information and communication technology. This is not a matter of heroic creative acts *ex nihilo*, however. The nature of the furniture added is shaped (enabled and constrained) by conditions, circumstances, and contexts, and the dynamics of this "shaping" contain a strong component of expectations and promises and how these influence action. I shall come back to the embedded nature of the new technological developments in the next section, but will here indicate the ontological implications of agenda dynamics.

As Van Lente argues in the final part of his (1993) thesis,

> This self-organizing character, this movement of an unreal projection converted into reality, is particularly striking in the case of technology. But technology only expresses something that is deeply human. As Ortega y Gasset [. . .] argued, projecting new forms of life and materializing them is the movement of human life itself.

He then goes on to quote Ortega:

> This invented life – invented as a novel or a play is invented – man calls "human life", well-being. [. . .] Have we heard right? Is human life in its most human dimension a work of fiction? Is man a sort of novelist of himself who conceives the fanciful figure of a personage with its unreal occupations and then, for the sake of converting it into reality, does all the things he does – and becomes an engineer? (Ortega y Gasset 1962, p. 108)

And he selects further quotes to support this point: The roots of all action is in the future, in the projections we design and consequentially try to convert into reality. "Body and soul are things, but I am a drama, if anything, an unending struggle to be what I have to be" (p. 113). "To live . . . that is to find means and ways for realizing the program we are" (p. 116). So, as a consequence: "Man has to be an engineer, no matter whether he is gifted for it or not" (p. 136).

I cannot resist quoting Van Lente's final paragraph:

> The reverse is true as well: the engineers have to be like novelists. They have to write forceful fiction and make it come true as well. Actors developing technology, to paraphrase Ortega, conceive for themselves the fanciful figure of a future technology, and for the sake of converting it into reality, they, as well as others drawn into the fancy, do all the things they do. Whatever technology may be in the present, it is rooted in the future.

THE WIDER WORLD OF TECHNOLOGY IN SOCIETY, AND THE FUZZY BOUNDARIES WITH THE ENGINEERING WORLD

Actors, artifacts, and agendas are always embedded in evolving and interacting contexts. To address such multi-level dynamics, Rip and Kemp (1998) introduced three levels at which technical change is played out: novel configurations, regimes, and socio-technical landscapes. (I note that their conceptualization focuses on technical change as such, and does not include industry structures and other patterns in society which are equally important to understand the socio-technical landscape.) Innovation and technological regimes have been discussed already; the interesting addition is socio-technical landscape – another invisible hand creating gradients of force (like a physical landscape). The upper level in Figure 2 represents the *longue durée* (Braudel, 1972; Bertels, 1973), the sedimented landscape resulting from earlier actions and cumulating infrastructures, and which changes only slowly. Like regimes, the socio-technical landscape enables and constrains, but it is constituted differently.

An example of analysis with the help of socio-technical landscape is the infrastructure of electricity generation and use, including networks and billing systems. This is how electricity has become embedded in society, and how its use has become so important that it is an almost obligatory passage point for fuels and other energy carriers in order to reach end users. In other words, there is a "buffer zone" in the socio-technical landscape that separates ongoing innovation (variation) in technologies to generate and distribute electricity, from innovations (variation) at the distribution and use side. Not only is selection constrained – in quasi-evolutionary terminology, by a nexus (Van den Belt & Rip, 1987), only now at the macro-level – but the range of variation is contained as well because it must be always lead to, or utilize, electricity. Alternatives to electricity stand little chance, at least in the short term.

The transport and mobility regime, centered on automobiles, is much more heterogeneous than the electricity regime, and could perhaps better be seen as a set of overlapping regimes. These are tied together by the reliance on vehicles using internal combustion engines, and what this implies for fuel, construction, maintenance, and ways of using automobiles. Staudenmaier (1989) has shown how this socio-technical regime became dominant and disciplined various communities. One can see a buffer zone again, less homogeneous than in the electricity sector, but still enabling further innovation while constraining it in particular directions. The development (and to some extent non-development) of electric vehicles (Hoogma, 2000) and of fuel cells (Schaeffer, 1998) are instructive examples.

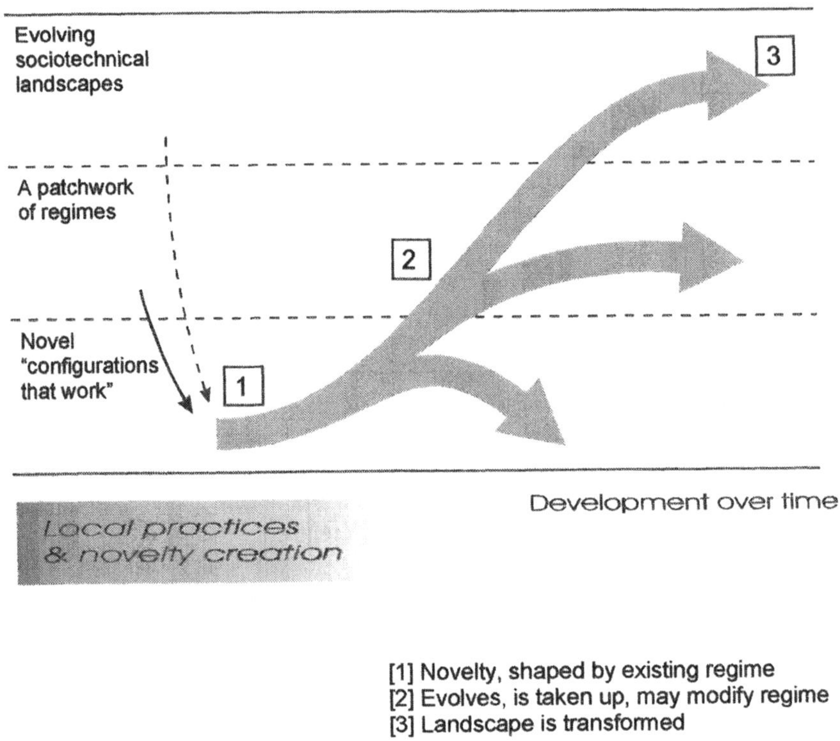

Figure 2. The three levels of socio-technical change

There are other meso- and macro-level dynamics, for example the shift from economies of scale to economies of scope (and now also economies of skill) as the joint outcome of industrial and technological developments (Chandler, 1990). In another vein, Hughes (1989) analyzed what he called the tidal wave of technological ingenuity and enthusiasm in the USA, which created a particular form of modernity (with hierarchical control orientation and tightly coupled systems), which is less appropriate to the present-day world and is being changed partly through technological developments in the direction of distributed systems (cf. Hughes, 1998).

Freeman (e.g., Freeman & Perez, 1988; Freeman, 1992) has introduced the notion of a techno-economic paradigm to capture the effect of what he calls a pervasive technology, that is a technology which not only changes its own sector but the whole economy because of pervasive effects in many sectors. One can argue that steam power, coupled with iron and steel, constituted the techno-economic paradigm of the railway (and steamship) age. While such a paradigm is dominant, technologies (in this case, electricity) develop that will characterize a subsequent paradigm. In retrospect, one can speak of a mismatch of the new technologies and the socio-institutional context shaped by the dominant techno-economic paradigm. For the present period, Freeman sees the new information and communication technologies as the emerging techno-economic paradigm of the 1990s and later decades. This is not so much a question of wealth creation (witness the productivity paradox), but of changing structures and interactions. One example would be the new possibilities for co-production when information exchange is not limited by geographical distance.

One can look at information and communication in another way as well, and study it as part of a broader secular trend: the increasing role of "software" in handling "hardware." Software development and use in computer technology is the obvious example, with the advent of programming languages in the late 1950s as the key step. Using the concept of software more broadly, one can see the advent of operations research, of "traffic engineering" (in telephone networks) and the increasing importance of logistics as examples of software. The design and "disciplining" of activities and organizations on the basis of blueprints are also software in the broad sense. Just like computer software, this generalized software is engineered, and thus qualifies as technology.

The broader notion of software links up with the analysis of historians and sociologists of how people, organizations, and society are monitored and disciplined with the help of technology. Michel Foucault has emphasized the "normalizing" tendency in modern societies, and the way technology is implicated in it (Foucault, 1975). Prisons are built in such a way that surveillance is optimal. At first, such technologies are specific to the particular purpose. But surveillance becomes a generalized problem: in hospitals, in armies, in shopping malls. And the engineering challenge shifts from the particular circumstance (the hardware) to ways to discipline people (the software).

Interestingly, the process of normalizing can also be seen in how the idea of what are "good" technologies becomes articulated – environmental technologies are good by definition, biotechnologies were OK until ambivalencies crept in and food ingredients deriving from genetically modified organisms became dubbed as "Frankenstein foods". One indicator of normalization is the importance of expectations and symbolic features: "no dirty chlorine in my food, or in the toys for my child, no products linked to genetically modified organisms in the shop." In other words, there are requirements about symbolic features which may dominate performance requirements. In these examples, negative associations were dominant, but positive connotations occur just as well, as with vitamins (Rip, 1991).

It is within these wider worlds that the engineers work and accommodate to their various (and changing) contexts. Their charter, or mainstay, is double: their products have a life of their own, as captured in Latour (1987)'s beautiful phrase of artifacts as "immutable mobiles"; and they work within a mandate where they are responsible for technological progress (Van Lente, 1993; Van Lente & Rip, 1998). Thus, a division of labor, and of responsibilities, is possible between engineers and the wider world.

Such boundaries and distinctions are not unproblematic, however, and may be called into question. A thought-provoking example are cochlear implants (Ragud & Ahlstrom, 1997). A double dynamic was visible: insiders (developers) were confronted by outsiders (regulators, providers, prospective users); and the market projections from the "hearing majority" met with strong reactions from the "deaf community" who see cochlear implants as an attempt to undermine deaf communication and culture.

This example indicates that the universalistic tendencies implied by the notion of technological effectivity (configurations that will work, or must work, wherever and for whomever) run up against what Staudenmaier called "human pluralism" (Staudenmaier, 1992, p. 224). An obvious, and often voiced, response is to call for participation, but this is only one response, and itself plagued by difficulties. Difficulties deriving from the general issue of ongoing practices, their partial rationalities, heterogeneous interactions and temporary closures (Rip, 1988; Rip, 1999), and from the multiple ontologies involved (Mol, 1999).

Engineers have been adept in closing off their rationalities from the wider world: such closure, which enables them to go on with present concerns, is their overriding goal. The call for opening up their practices to participation, however well intentioned, is not a sufficient reason by itself. Citizen groups have their own partial rationalities, for example NIMBY – Not In My Back Yard. Discussion of boundaries and participation should be framed by considerations of short-term and long-term quality of technology in society. To do so, one needs (in the end) a technological version of Kant's *Urteilskraft*. Judgmental competence has to be built up and valued. And tools to support judgments, for example reflective scenarios, have to be developed.

Increasing interest in participation, in giving voice to those who used to be excluded from engineering/technological decisions, has led to short-circuiting the quality issue by assuming that wider participation, somehow, leads to better technology (in a better society). This is doubtful, to say the least, and work on Constructive Technology Assessment (Rip et al., 1995; Schot & Rip, 1997) has identified approaches that do not depend on participation by spokespersons, and has studied concrete cases to evaluate possibilities and limitations on this count.

By contrast, one should recognize that in emphasizing quality assurance and the eventual goal of better technology in a better society, other considerations, in particular issues of democracy and justice, are backgrounded. At a minimal level, the requirement of transparency, which is important in a democracy (Van den Daele, Pühler & Sukopp, 1997, p. 90), may not always be conducive to productive negotiation. More generally, broad participation, while perhaps a "right," is not a productive way to encompass variety. Van den Daele, who has been confronted with these issues repeatedly, suggests that the multiplication of viewpoints (as a result of increased participation) will make the achievement of an integrated result and formulation of concrete policies more difficult. This could lead to a return to formal democratic policy and decision making procedures, unless civic society itself is able to achieve such integration (Van den Daele & Neidhart, 1996, p. 14). Clearly, this will not come about by itself.

In the reflexive co-evolution of technology and society, dominant patterns, and thus dominant designs, emerge that allow for such integration. The proponent-opponent dichotomy and the risk discourse which emerged in the 1960s and 1970s is one example of such a dominant *de facto* design. One may have one's doubts as to its productivity, especially in the face of the new technologies and the new circumstances of the 1990s, but it is definitely a solution to Van den Daele's concern.

When discussing these issues (Rip, 1999), I also asked what might be (and could actually become) a dominant design for managing technology in society, better suited to the times and technologies? One can specify general "societal" design requirements: It must always produce kaleidoscopic integration, that is, integration without doing away with antagonisms and outlyers. It must relate to ongoing societal dynamics and accept the pragmatic limitations this implies: to have an ideal approach which does not work because it does not fit the tribal norms is of little purpose. Engineering design can then be positioned as part (and parcel) of societal design. Backgrounding the wider aspects can be productive in the short term, but risks having them return with a vengeance. For modern biotechnology, this is now happening. And for infrastructural investments, a similar backlash is possible.

ISSUES FOR AN EMPIRICALLY-ORIENTED PHILOSOPHY OF TECHNOLOGY

There are many explicit and implicit philosophical issues embedded in what has been said, some of which, have been discussed. In this concluding section

I shall briefly focus on two issues: the methodology of the empirical turn, and the political ontology of engineering.

How to learn from phenomena? The principle problem is that one has to have a *prior* diagnosis of the interest and value of the case at hand to select it as an occasion for learning – but then, the original aim was to learn from what one encounters, rather than selecting beforehand . . . . Carl Mitcham, when confronted with this challenge (in discussion after a lecture at the Technological University of Delft), offered a solution: learn here, in case A, and use these insights to advise and improve what happens elsewhere (in cases B, C, . . .). If the analyst has sufficient grounds to diagnose what is "good", this route can be followed even if path dependencies can occur which introduce limitations (Thagard, 1988; cf. also Lipton, 1993). Recognizing the limitations that path dependencies will introduce is a key step to start an open-ended learning process.

What are sufficient grounds to identify which cases are "good", or at least exemplary? We are confronted here with the general problem (or challenge) of naturalistic approaches: in order to learn about "good" science or "good" technology, one has to select cases as exemplifying good science or technology, and thus have some prior criterion about what is good and what is not. Thagard (1988) has addressed this issue, and shown that the vicious circle can be turned into a virtuous circle by learning as one goes along. (He also shows that the narrow reflective equilibrium that can be achieved in this way has to be evaluated in terms of a broad equilibrium.)

Donald Schön's (1983) analysis of learning by practitioners is helpful as well. Engineers, architects, and psycho-analysts all learn in games with the situation, and transfer their insights by considering new situations as instances of types of situation they have encountered before. Similarly, philosophers and sociologists of technology can play a game with their case-studies, by naming them *as case of*, say, bottom-up aggregation, and try out whether their data can be fitted into the story they want to tell, in partial contrast to other analysts' stories. Clearly, this is not a recipe, but a heuristic that enables learning.

Political ontology (of engineering) is a new phrase, coined for this occasion. It does resonate with Mol's (1999) discussion of "ontological politics", and her discussion of "multiple realities" which can be, and are, performed – which includes partial materialization. As she emphasizes, reality is not an immutable given for technologists and politicians. The point of technology and of politics is

> the assumption that the world might be mastered, changed, controlled. So within the conventions of technology and politics the question of how to shape reality was open: at some point in the future it might be otherwise.

And she adds: "But along with this it was assumed that the building blocks of reality were permanent: they could be uncovered by sound scientific investigation.") What I add here is the recognition that engineers, in interaction with their contexts, are attempting to reduce the "multiple realities" that are projected, to one dominant ontology, in the sense of the actual furniture of the world. It is not just the multiplicity that gives rise to politics (even if such reconstructions are important to convince the Horatios of this world that there is more between heaven and earth than originally thought), but the negotiations and concrete assemblage work which reduce multiplicity to a concrete configuration.

In addition to the materialization of stories, and the issues of quality and participation, which I have discussed already in the previous section, there is an underlying issue that has to do, in Hannah Arendt's words, with our responsibility for a world which we create ourselves (Arendt, 1958, pp. 136–138; Achterhuis, 1999, pp. 11, 116–117). As Achterhuis emphasizes, this is not about the morality of our actions and good intentions toward our fellow human beings, but about the political question of establishing and maintaining arrangements in and of the world. Engineers are well placed to take up this responsibility (and are actively involved in establishing and maintaining the arrangements of the world), but rarely have the competencies to address the issues of political ontology productively.

Traditionally, private spaces have been the topic of philosophy (cf. Heller's 1987 critique), but there is in-between space, and its cumulation to public space. This is what Arendt calls "the world", and care for the world is an attempt to maintain in-between space, with its tensions. Public space is supported, and shaped by material infrastructures, which themselves derive from actions and interactions. Political action and interaction is thus continuous with material shaping. Conversely, the engineers' work adds to the material landscape, and thus shares with political action the care for the world.

Thus, design of public places – Schiphol airport is a very interesting example – , and of infrastructures, are strategic sites for an empirically-turned philosophy of technology. This is not just because of the presently visible controversies – these are fought at a surface level – , but their occurrence is an indication of a struggle about care for the world.

What I have tried to show here is that the empirical turn is visible and viable, and that it promises a better philosophy of technology. Accepting the courage implied in the convictions set out here, I add that my reflections also presage a better philosophy in general. But that is another story.

REFERENCES

Achterhuis, H. (1999). *Politiek van goede bedoelingen.* Amsterdam: Boom.

Achterhuis, H., Smits, R., Geurts, J., Rip, A., & Roelofs, E. (Eds.) (1995). *Technologie en Samenleving* A Coursebook of the Dutch Open University, Heerlen. Leuven/Apeldoorn: Garant.

Arendt, H. (1958). *The Human Condition.* Chicago: University of Chicago Press.

Bertels, K. (1973). *Geschiedenis tussen struktuur en evenement.* Amsterdam: Wetenschappelijke Uitgeverij.

Braudel, F. (1972). *The Mediterranean and the Mediterranean World in the Age of Philip II.* London: William Collins. (Translation of the second revised French edition, 1966).

Brown, R. H. (1989). *A Poetic for Sociology: Toward a Logic of Discovery for the Human Sciences.* Chicago: University of Chicago Press.

Chandler, A. D. Jr. (1990). *Scale and Scope: The Dynamics of Industrial Capitalism.* Cambridge, MA: Harvard University Press.

Foucault, M. (1975). *Surveiller et Punir: Naissance de la Prison.* Paris: Gallimard.

Freeman, C. (1992). *The Economics of Hope: Essays on Technical Change, Economic Growth and the Environment.* London: Pinter.

Freeman, C., & Perez, C. (1998). Structural Crises of Adjustment, Business Cycles and Investment Behaviour. In: G. Dosi, et al. (Eds.), *Technical Change and Economic Theory.* (pp. 38–66). London: Frances Pinter.

Garud, R., & Ahlstrom, D. (1997). Technology Assessment: A Socio-cognitive Perspective. *Journal of Engineering and Technology Management 14* 25–48.

Heller, A. (1987). Sociology as the Defetishation of Modernity. *International Sociology* 2(4) 391–401.

Hoogma, R. (2000). *Exploiting Technological Niches. Strategies for Experimental Introduction of Electric Vehicles.* Ph.D. Thesis, University of Twente, 26 May.

Hughes, T. P. (1989). *American Genesis: A Century of Invention and Technological Enthusiasm, 1870–1970,* New York, Viking Press.

Hughes, T. P. (1998). *Rescuing Prometheus.* New York: Pantheon Books.

Latour, B. (1987). *Science in Action.* Milton Keynes: Open University Press.

Lipton, P. (1993). *Inference to the Best Explanation.* London: Routledge.

Mol, A. (1999). Ontological Politics: A Word and Some Questions. In: J. Law & J. Hassard (Eds.), *Actor Network Theory and After* (pp. 74–89). Oxford: Blackwells.

Morley, P. (Member of the Board, KPN Telecom) (2000). Lecture, University of Twente, 7 February.

Nelson, R. R., & Winter, S. G. (1977). In Search of a Useful Theory of Innovation. *Research Policy 6* 47–76.

Ortega Gasset, Jose (1962). Man the Technician. In: *idem, History as a System* (pp. 87–164). New York: Norton. (Original text from 1940).

Ravetz, J. (1975). '. . . et augebitur scientia'. In: R. Harré, (Ed.), *Problems of Scientific Revolution. Progress and Obstacles to Progress in the Sciences* (pp. 42–57). London: Oxford University Press.

Rip, A. (1988). *Technologie als Mensenwerk.* Inaugural Lecture, Enschede, Universiteit Twente, 13 Oct.

Rip, A. (1991). The Danger Culture of Industrial Society. In: R. E. Kasperson, & P. J. M. Stallen (Eds.), *Communicating Risks to the Public: International Perspectives* (pp. 345–365). Stallen. Dordrecht: Kluwer Academic.

Rip, A. (1999). *Contributions from Social Studies of Science and Constructive Technology Assessment.* ESTO Project on Technological Risk and the Management of Uncertainty, commissioned by the Forward Studies Unit, European Commission. Enschede: University of Twente, April.

Rip, A., & Kemp, R. (1998). Technological Change. In: S. Rayner & E. L. Malone (Eds.), *Human Choice and Climate Change* (vol. 2, ch. 6, pp. 327–399). Columbus, Ohio: Battelle Press.

Rip, A., Misa, T. J., & Schot, J. (Eds.) (1995). *Managing Technology in Society: The Approach of Constructive Technology Assessment, Constructive Technology Assessment.* London and New York: Pinter.

Rip, A., & Schot, J. (1999). Anticipating on Contextualization – Loci for Influencing the Dynamics of Technological Development. In: D. Sauer, & C. Lang (Eds.), *Paradoxien der Innovation – Perspektiven sozialwissenschaftlicher Innovationsforschung, sozialwissenschaftlicher Innovationsforschung* (pp. 129–146). Frankfurt/New York: Campus Verlag.

Schaeffer, G. J. (1998). *Fuel Cells for the Future: A Contribution to Technology Forecasting from a Technology Dynamics Perspective.* Ph.D. Thesis, University of Twente, 6 November.

Schön, D. A. (1983). *The Reflective Practitioner: How Professionals Think in Action.* New York: Basic Books.

Schot, J., & Rip, A. (1997). The Past and Future of Constructive Technology Assessment. *Technological Forecasting and Social Change, 54* 251–268.

Science. (1996). Can Chip Devices Keep Shrinking? 13 December, 1834.

Staudenmaier, J. M. (1989). The Politics of Successful Technologies. In: R. C. Post & S. H. Cutliffe (Eds.), *In Context: History and the History of Technology. Essays in Honor of Melvin Kranzberg* (pp. 150–171). Bethlehem, PA: LeHigh University Press.

Staudenmaier, J. M. (1992). Science and Technology: Who Gets A Say? In: P. Kroes & M. Bakker (Eds.), *Technological Development and Science in the Industrial Age* (pp. 205–230). Dordrecht: Kluwer.

Thagard, P. (1988). From the Descriptive to the Normative. *Computational Philosophy of Science.* (ch. 7, pp. 113–137). Cambridge, MA: MIT Press.

Van Lente, H. (1993). *Promising Technology – The Dynamics of Expectations in Technological Developments.* Ph.D. thesis, University of Twente. 25 November.

Van Lente, H., & Arie Rip, A. (1992). *Construction by Expectation: An Essential Ingredient of Technology Dynamics.* Paper presented at the 4S/EASST Conference, Gothenburg (Sweden) August 12–15.

Van Lente, H., & Rip, A. (1998). Expectations in Technological Developments: An example of Prospective Structures to be Filled in by Agency. In: C. Disco & B. J. R. van der Meulen (Eds.), *Getting New Technologies Together (pp.* 195–220). Berlin: Walter de Gruyter.

Van de Poel, I. (1998). *Changing Technologies: A Comparative Study of Eight Processes of Transformation of Technological Regimes.* Ph.D. thesis, Twente, Enschede: University Press.

Van den Belt, H., & Rip, A. (1987). The Nelson-Winter/Dosi Model and Synthetic dye chemistry. In: W. E. Bijker, T. P. Hughes & T. J. Pinch (Eds). *The Social Construction of Technological Systems: New Directions in the Sociology and History of Technology* (pp. 135–158). Cambridge, MA: MIT Press.

Van den Daele, W., & Neidhardt, F. (1996). Regierung durch Diskussion – Über Versuche, mit Argumenten Politik zu machen. In: W. van den Daele, F. Neidhart, (Eds.), *Kommunkation und Entscheidung: Politische Funktionen öffentlicher Meinungsbildung und diskursiver Verfahren* (pp. 9–50). Berlin: Edition Sigma, WZB-Jahrbuch.

Van den Daele, W., Pühler, A., & Sukopp, H. (1997). *Transgenic Herbicide-Resistant Crops: A Particpatory Technology Assessment. Summary Report.* Berlin: Wissenschaftszentrum Berlin für Sozialforschung.

ENGINEERING DESIGN AND THE EMPIRICAL TURN IN THE PHILOSOPHY OF TECHNOLOGY

Peter Kroes

1. PHILOSOPHY OF TECHNOLOGY: ITS AIM AND RELEVANCE

In my opinion, one of the principal aims of philosophy in general is *conceptual clarification*, that is, clarification, on the basis of *critical* reflection or analysis, of basic notions with the help of which we describe human experience. In line with this idea, I take the *aim* of philosophy of technology to be a critical reflection on or analysis of the concepts and conceptual frameworks with the help of which we formulate and express our thinking about technology and topics related to technology, such as the effect of technology on human life and society.

The possible *relevance* of such a conceptual clarification, and therefore of the philosophy of technology, is of a twofold nature, i.e., intellectual and pragmatic. On the one hand, technology is an interesting topic for philosophical reflection in its own right; that is, intellectual curiosity may be its main driving force. Apart from any pragmatic considerations, conceptual problems related to technology – for instance, the distinction between natural objects and artifacts – may be of interest from a purely intellectual point of view (the need to understand the meaning of basic categories in terms of which we frame our thinking about the

Parts of this article appeared previously in Techne, http://scholar.lib.vt.edu/ejournals/SPT Vol 3, No. 3, Spring 1998 and is used here by permission.

The Empirical Turn in the Philosophy of Technology, Volume 20, pages 19–43.

ISBN: 0-7623-0755-2

world). On the other hand, there is a widespread concern about the impact of technology on human beings and society, which gives rise to ongoing debates and controversies about technology. These debates often have a direct pragmatic dimension, in the sense that they center on concrete decisions to be taken about realizing new technological possibilities (e.g., cloning of humans). But they also address more general issues with a more diffuse pragmatic nature, such as the control of technological development as a whole. In these debates, various images of or presuppositions about technology (e.g., technology as an autonomous force) play a role. The philosophy of technology may contribute to a clarification of these images and presuppositions, and thereby prove to be of value for these pragmatic issues. As Elisabeth Ströker remarks (1983, pp. 323–336):

> It must not be overlooked, however, that every practical problem has a theoretical dimension which cannot be overlooked. It could turn out, here as elsewhere, that downplaying theoretical questions about technology has its risks; neglect of philosophical discussion might hide contradictions and prevent equivocations from becoming apparent, and erroneous arguments might pass unnoticed. Thus the discussion of the pragmatic problems of technology would in the long run be affected too.

The broader universe of discourse in which pragmatic issues about technology are discussed, raises conceptual (theoretical) issues, and the way these issues are resolved may affect the way in which the pragmatic issues themselves are resolved.

Given the above statements concerning the aim and relevance of the philosophy of technology, and given that technology pervades almost every aspect of human life, that the human-made world plays a crucial role in human experience, and that the influence of modern technology on human existence and society is still increasing and is raising many issues, one would expect the philosophy of technology to be a thriving field. However, nothing is less true. Within philosophy, the philosophy of technology is only a marginal area of interest. But also outside that context, the philosophy of technology plays a marginal role, e.g., in engineering curricula and in discussions in society at large that touch on technology. The philosophy of technology simply does not matter! According to Albert Borgmann, this is the case not only for the philosophy of technology, but for philosophy in general. In his article "Does Philosophy Matter?" (1995) he complains that philosophers are invisible in American intellectual life. Nevertheless, modern philosophy is, in his opinion, facing a challenging task, namely "the radical and normative examination of the material culture that has been shaped by modern technology and is about to be transformed once again by the vigor and ingenuity of recent technological developments" (Borgmann, 1995, p. 295). Indeed, a challenging task for philosophy in general, a task to which the philosophy of technology could and should contribute, but in fact does not.

Why does the current philosophy of technology play such a marginal role in comparison to, for instance, the philosophy of science, and more generally, why is it almost invisible inside and outside academic circles?[1] Although the diagnosis that the philosophy of technology suffers from insignificance may be easy to make, the unraveling of the reasons for this state of affairs is more difficult. Surely, part of the imbalance in the interest in the philosophy of science and the philosophy of technology may be due to the difference in status of science and technology in Western culture, a difference inherited from Greek antiquity. But this imbalance is also partly due, in my opinion, to specific features of the philosophy of technology itself. In the following section, some of these features will be discussed briefly.

2. TRADITIONAL PHILOSOPHY OF TECHNOLOGY: A MISNOMER

In the first place, it must be noted that the philosophy of technology can hardly be called a philosophical discipline, from either an institutional or a substantive point of view. The Society for Philosophy and Technology (SPT) is the only significant society; it organizes conferences on a regular, biennial basis, and its electronic journal is the only international journal in this field.[2] This is about all the institutional infrastructure there is on an international level for the philosophy of technology.

From a substantive point of view, the philosophy of technology lacks coherence. Presumably, the above statement concerning the aim of the philosophy of technology will not provoke much controversy among philosophers interested in the phenomenon of technology. I suppose most of them will feel comfortable about it, because it is so broad that it encompasses most of the work done in the field traditionally called the "philosophy of technology". But that is as far as their agreement will go. Controversies arise as soon as it is asked how the critical reflection or analysis is to be performed, what the subject matter of this critical reflection (i.e., technology) is, and what questions are central to the field. Here, opinions start to diverge, leading to various approaches within the philosophy of technology, such as the anthropological, phenomenological, social-philosophical, Anglo-Saxon analytical, and continental speculative approaches, each addressing in their own way what they take to be the central issues in the philosophy of technology. Representatives of these different approaches often find little common ground for discussion.[3] As a result, this branch of philosophy is a strongly fragmented field of research: There is no consensus about its subject matter, or the methods by which to deal with it.[4] The field suffers from what Ströker calls the "paradox of its continual beginning"; looking at the philosophy of technology, "one is still confronted at present only with a multitude of detailed philosophical studies each of which can be characterized as merely a sketch or an attempt" (1983, pp. 323–324).

The most striking common feature of much work done in the philosophy of technology, however fragmented this field may be, is the externalistic approach to technology. What goes on in the world of technology and engineering itself – i.e., the design, development, production, and maintenance of technological objects – remains out of sight. Technology itself (however understood: as a collection of artifacts, a form of knowledge, a form of human action, a social process, etc.) usually remains a black box, and often technology is treated as an undifferentiated whole (this is especially the case in speculative/metaphysical approaches). There is a strong focus on philosophical problems related to the *consequences* of the use of technological artifacts for human beings, society, and culture, and on moral problems involved in the use of technology, not on philosophical (moral) problems related to the making of technological artifacts. In my opinion, this is one of the main reasons why the philosophy of technology suffers from insignificance; this branch of philosophy is not taken seriously because it itself does not take technology seriously.

Let me illustrate the neglect of what goes on in the world of technology and engineering in more detail. I cannot but share Joseph Pitt's frustration when, discussing Frederick Ferré's book *Philosophy of Technology*, Pitt (1990, p. 13) writes:

> How can you write a book about the philosophy of technology and not discuss in detail either engineering or science? How can you write a book about the philosophy of technology and worry primarily about religion and metaphysics? These questions have troubled me deeply after having studied Ferré's *Philosophy of Technology* with some care.

I had similar feelings after reading Don Ihde's *Philosophy of Technology: An Introduction* (1993). Note that this introduction to the philosophy of technology contains in the index the following two entries concerning design: Designer intent 116, and Designing and technology 48. Two references to design, that is all! Nowhere in this *introduction* to the philosophy of technology is there to be found a systematic treatment of one of the core activities of technology, namely designing. As if there are no fundamental philosophical issues involved in the notion of design!

Mitcham's book *Thinking through Technology* raises mixed feelings. I consider this book one of the best introductions to the philosophy of technology – not only because it offers a good, systematic survey of the philosophy of technology, but also because he stresses that philosophers should pay more attention "to what really goes on in engineering and technology" (1994, p. ix). I fully endorse this point of view. At the end of the book, Mitcham writes that it was his intention

> to take technology seriously, philosophically seriously. The chapters of Part 2 thus intentionally wallow in the details of engineering texts (p. 268).

Unfortunately, however, the search for those details in Part 2 is in vain. Mitcham does not discuss in detail any concrete example of technology (how it was designed, developed, implemented, what kind of problems had to be solved, etc.) in order to formulate and illustrate philosophical issues related to technology.[5]

When we compare studies in the philosophy of technology with those in the philosophy of science, the most striking feature of the former is the distance between the philosophy of technology and technology itself; the philosophical examination of technology seldom is performed in terms of concrete examples from technology. It seems that technology itself, in particular the design and development of technology, is a field where "few philosophers dare to tread" (Ihde, 1993, p. 140). In view of the foregoing, the expression "philosophy of technology" is simply a misnomer for much that goes on in this field.

To avoid misunderstanding, let me explicitly state that it is not my intention to argue that philosophical issues concerning the consequences of the use of technology are not important for, or do not belong to, the field of the philosophy of technology. On the contrary, the philosophy of technology has to address these questions, as it is here (but not only here) that it may show its pragmatic relevance. What I am arguing for is that a fruitful philosophical discussion of many of these issues requires at least *some* knowledge of what goes on inside technology and engineering. In this respect, I completely agree with Mitcham when he observes that "it is remarkable, for instance, that none of the standard texts in engineering ethics contains any serious analysis of the engineering process as such" (1994, p. 7). He puts forward the suggestion that fundamental theoretical issues about technology are relevant for dealing with ethical issues in engineering, and that in this way "it may ultimately be possible to make a more profound contribution to ethical reflection than by immediate engagement with particular moral problems" (ibid). I think Mitcham's point with regard to engineering ethics is valid for philosophical reflection on technology in general. Many of the consequences of technology cannot be properly analyzed and understood philosophically without taking into account the way technological artifacts are designed, produced, and implemented in social practices. This is especially true for so-called socio-technical systems (such as transportation systems), where there is a close interlocking of social and physical aspects. The extent to which the black box of technology has to be opened in philosophical reflections, depends, of course, on the specific case at hand.

It is also not my intention to reject offhand as irrelevant to the philosophy of technology, philosophical (metaphysical) inquiries into the nature of technology. But here, more or less the same is true as for philosophical reflections on the consequences of technology. Just as the nature of science, as an institutional and as an intellectual enterprise, underwent dramatic changes over the course of time, so has the nature of technology. How is it then possible to conduct a metaphysical analysis of the nature of technology without taking into account such changes?

Apart from these critical remarks on much of the work done in philosophy of technology up till now, there is another point that I would like to make, namely that technology itself – i.e., the creation of technological artifacts – raises interesting philosophical problems. In his discussion with Ferré, Pitt (1990, p. 4) makes a similar point.

> [P]hilosophical questions of interest concerning technology center around the decisions we make regarding technological enterprises. By "technological enterprises" I mean such diverse projects as the building of a nuclear plant, the development of a new weapon system, the creation of new consumer goods and their marketing, the creation of social institutions and the development of knowledge. The main problems concern the nature of the knowledge and the values we use to make decisions about the development and implementation of projects in these areas. In these contexts major questions arise concerning the kinds of risks which might be incurred and the reliability of techniques for correctly estimating those risks as well as other consequences, intended or unintended.

I agree with Pitt that decisions made regarding technological enterprises prompt interesting philosophical questions concerning the nature of knowledge and values, and that often these problems have received little attention in the philosophy of technology.[6] But there is yet another topic for philosophical reflection. Although not mentioned explicitly by Pitt, it belongs at the center of any philosophy of technology that claims to take technology itself seriously, namely the nature of technological (engineering) design. Designing is one of the core activities in technology (engineering), and it raises many interesting philosophical problems which have been grossly neglected by philosophers of technology. I will now turn to a brief survey of some of these problems.

3. EXPLORATION OF A PHILOSOPHY OF TECHNOLOGICAL DESIGN

One of the most persistent myths about technology is that it is applied science (e.g., the transistor is an application of quantum mechanics, the radio of the theory of electromagnetism, the Haber-Bosch nitrogen fixation process of thermodynamics, etc.). According to this myth, technological knowledge is a special form of scientific knowledge: It is scientific knowledge applied to situations that are technologically relevant. I assume that this view of technology is one of the reasons why philosophers of science have never seriously studied technological knowledge. This myth also suggests that it is somehow possible to derive designs for technological devices from scientific theories (knowledge): Add the right boundary conditions to those theories, and the designs will come out. This myth has been rejected time and again as inadequate by philosophers of science and technology. But the single most important reason why it leads to an inadequate picture of technology is that this view shows a complete lack of appreciation for the specific nature of a technological design. The notion of design is of no real importance in the physical sciences, except in a derivative sense, e.g., when physicists speak of the design of an experiment or of designing a theory. But take away the notion of design from technology and you can no longer make any sense of technology itself. Generally speaking, however, philosophers have failed to make technological design, and therefore technology, a topic of systematic philosophical reflection.

When we try to catch sight of technological design from a philosophical perspective, four related clusters of problems present themselves, labeled here *foundational, epistemological, methodological,* and *ethical* issues. What follows is a brief description of these clusters.

(1) Foundational Issues

The term "technology" is used with a variety of meanings (as a form of knowledge, a collection of artifacts, a form of human action, etc.), each of which brings forth fundamental issues. For example, when technology is taken in the sense of a collection of artifacts, it raises the question how to characterize technological artifacts as distinct from natural objects. This is not only a conceptual problem, but it also has pragmatic ramifications. The distinction between the artificial and the natural leads to problems concerning the relation between humans and nature; however, it also plays, for instance, a role in discussions about the patentability of genetically modified organisms (think of Harvard's onco-mouse). This topic is intimately tied up with questions about the meaning (or meanings) of the notions of design and technological design. In what sense(s) are these notions used by engineers? A characteristic feature of the engineering description of objects, as contrasted with a physical description, is the use of a mixture of structural and functional concepts. The notion of technological design is a locus where the structural and functional descriptions of the world somehow meet. The notion of function puts the problem of teleology upon the stage. When the term "technology" refers to a specific form of human action, questions about the notion of technological rationality appear. Other foundational issues concern the meaning of specific notions used within a technological context, such as the notion of information.

(2) Epistemological issues

These involve, among other things, a characterization of the specific nature of technological (design) knowledge (containing descriptive and prescriptive elements) in comparison to, for instance, scientific knowledge, the formation of concepts in technology and the technological sciences, and the relation between knowledge and action (theory and praxis) in technology. Which types of technological knowledge may be distinguished? Technological knowledge ranges from engineering theories – in which knowledge is available in conceptualized form and often expressed with the help of formal languages – to practical skills, in which case knowledge is not available in conceptualized form. What are the criteria used for evaluating technological knowledge claims? Interesting questions with regard to technological design concern the different types of knowledge used by engineers in solving design problems. Related to these are issues about the relation between science and technology, the use and role of idealizations in science and technology, the nature and structure of technological explanations in which technological functions of artifacts are explained on the basis their (physical) structure, etc.

(3) Methodological issues

Many questions arise when we consider the designing of technological artifacts as a problem-solving process. What kinds of design problems may be distinguished (compare, for instance, the designing of a software program with the designing of a bridge)? Which methods are used for solving design problems, are these methods specific to technology in comparison to science, and how is the use of these methods justified? What is the role of heuristic rules in solving design problems, and how are heuristic rules justified (empirically, theoretically, the role of science)? Which criteria are used for evaluating solutions to design problems? What kind of constraints that designs have to satisfy may be distinguished (natural, technological, social)? How are conflicts between various types of constraints handled by engineers (tradeoffs between specifications)? How are functional requirements translated into technical specifications? How are very general social requirements (such as sustainability and safety) translated into concrete design specifications for products? How do engineers deal with risks due to various failure modes of technological artifacts, and how do they gather reliable information about (the acceptability of) risks? Etc.

(4) Ethical issues

In design processes decisions are made that have influence on the use/misuse and the social implementation of technological products. It is quite common for designers to have to make tradeoffs between, for instance, economic costs and risks (e.g., roll-on, roll-off ferries). This raises the issue of the moral and social responsibility of engineers involved in design processes, i.e., the issue of their professional responsibility. To what extent are design engineers morally responsible for the dangers, risks, and possible misuse of their technological products? This is one of the central issues of engineering ethics. In dealing with this problem, certain aspects of the professional practice of engineers have to be taken into account. One important aspect is the division of labor in design processes; designing a technological product is usually no longer done by an individual, but by teams composed of specialists from various fields. This raises the problem of the relation between individual and collective responsibility. Another important aspect of the professional practice of engineers is that they usually perform their work as employees of firms, which may severely restrict their freedom of action. To what extent does this affect their moral responsibility for the consequences of their actions?

Although there are scattered, valuable attempts to deal with some of the problems mentioned above, there is no sustained tradition in the philosophy of technology in which the design phase of technology is made the object of systematic philosophical reflection.

4. A THREEFOLD REORIENTATION

If my diagnosis of the cause of the present malaise in the philosophy of technology is correct, a threefold reorientation in this field appears to be necessary, namely with regard to the topics addressed, with regard to the level of analysis, and with regard to its method.

First, with regard to the topics addressed: I have argued above that a shift of attention from the user phase to the phase of design, development, and production of technological artifacts is not only desirable because technology itself raises many interesting philosophical issues, but also to make possible a fruitful treatment of many philosophical issues concerning the user phase of technological products.

Second, with respect to the level of analysis: Much work done in the philosophy of technology is performed at a very high level of abstraction at which technology is treated as some kind of undifferentiated whole. These often ahistoric and monolithic views on technology stand in sharp contrast to the variety of different forms of modern technology and engineering (mechanical, chemical, physical, architectural, genetic, software engineering, etc.) It is not at all self-evident whether there is a common element to these different forms, and if so, what this common element is. In this respect, it is worthwhile to recall Wittgenstein's well-known analysis of the notion of 'Spiel' in his *Philosophische Untersuchungen.* All the various forms of technology (engineering) may hang together through relations of family resemblance, without there being a common core element. Modern technology is a historically grown, highly complex and diverse phenomenon, a fact that should not be ignored by philosophers of technology. This is only possible through a shift from a global to a more local level of analysis. The richness of technology will become visible only by looking at modern technology through a magnifying glass.

Finally, in relation to method: A much more important role for empirical material in philosophical studies of technology appears to be necessary. Many of the issues mentioned above cannot be studied fruitfully without taking into account what 'really' goes on in technology, that is, without detailed empirical descriptions of concrete technological developments.

Of course, these three points hang together: A fruitful philosophical analysis of problems related to specific (local) developments in technology appears to be possible only on the basis of empirical descriptions of these developments. The following section contains an example to illustrate this "empirical approach" in the philosophy of technology. It discusses the problem of the relation between structure and function of technological artifacts in terms of a concrete example.

5. EXAMPLE: THE RELATION BETWEEN STRUCTURE AND FUNCTION IN TECHNOLOGICAL ARTIFACTS

A technological object such as a television set or screwdriver has a dual nature.[7] On the one hand, these are *physical objects* with a specific physical structure (physical properties), the behavior of which is governed by the laws of nature. On the other, an essential aspect of any technological object is its *function.* A technological object "has" a function, which means that within a context of human action it can be used as a means to an end. A physical object is the carrier of a function, and it is by virtue of its function that the object is a *technological* object. Usually a technological object is the embodiment of a human design and is specifically made to realize a certain function. Function and physical carrier together constitute a technological object. The function cannot be isolated from the context of use of a technological object; it is defined within that context. Since that context is a context of human action, we will call the function a human (or social) construction. Thus, a technological object is a physical construction as well as a human/social construction.

This dual nature of technological objects is reflected in two different modes of description, viz. *a structural* and a *functional mode*. Insofar as it is a physical object, a technological object can be described in terms of its physical (structural) properties and behavior. This structural mode of description makes use of concepts from physical laws and theories, and is free of any reference to the function of the object. The language of modern physics has no place for functions, goals, or intentions. With regard to its function, a technological object is described in an intentional (teleological) way: The function of a television set is to produce moving pictures, and that of a screwdriver to tighten or loosen screws (Searle, 1995). Purely functional descriptions of an object have, from a structural point of view, a black box character in the sense that they do not specify any physical properties of the object: A television set is *something* (whatever it may be) which produces moving pictures, and a screwdriver is *something* with which to tighten and loosen screws.

A closer look at the notions of design and design process also reveals the dual nature of technological objects. Here, the notion of design is taken primarily in the sense of a complete description of a physical object, such that on the basis of that description an object with a particular function can be made. A design is the outcome of a design process, which typically starts by defining the commercial requirements specification of an artifact; the reason for doing so may be an explicit market demand or a *technological* opportunity that has arisen. At this stage, the new artifact is defined in terms of its *global functional properties and costs*. This stage is usually followed by an inquiry into the technical and commercial feasibility of the artifact. After several other intermediary steps,[8] the design process ends with a complete detailed *technical/physical* characterization of the artifact, which contains a description of the technical/ physical specifications of the artifact as a whole and of each and every part of it. In other words, in the course of the design process *the function to be performed has been translated into a construction to be produced*. The final step is the technical validation/verification of a design; this roughly consists of showing that a real prototype version of the artifact, which until then existed only on paper, indeed adequately performs its intended function.

Thus, the design process typically starts with the description of a required function: Something to produce moving pictures, or something to tighten or loosen screws. As already remarked, a characteristic feature of functional descriptions is that they do not refer to any physical properties of the desired object. The starting point of a design process is therefore a kind of *black box* characterization of a physical system. The task of the designer is to show how the *black box* may be filled with a physical system, such that the required function will be performed by that system. The proposed design should contain a complete description of that physical system. It is important to note that a design is more than just a complete description of the physical properties of a thing to be made: It also contains (at least implicitly) an explanation of how the proposed physical system will be able to perform the required function. In other words, a design also consists of a *technological explanation*, i.e., *an explanation of the function of a technological object* in terms of the physical structure of that object. A technological explanation is an integral part of a design, and plays a crucial role in justifying a design: It shows that on the basis of its physical structure, an object will perform a certain function.

The claim that an adequate design contains a technological explanation raises a fundamental problem. It means that engineers are somehow able to bridge the gap between a structural and a functional description of a technological object: A function – which is described in an intentional language – is explained in terms of a structure, which is described in a non-intentional, structural language. How is this possible? What does this mean for the relation between the structural and functional modes of describing technological objects? Can one mode of description be reduced to the other? Is it possible to deduce from a complete structural (scientific) description of an object its technological function, or vice versa? The answer to the last question is usually negative.

For instance, the historian and philosopher of technology Walter Vincenti claims, following Polanyi, that the operational principles of technological objects are not already contained in the laws of nature. According to Vincenti and Polanyi, operational principles – which describe how a technological artifact fulfills its function – cannot be deduced from those laws (Vincenti, 1990, p. 209):

> Finally, the operational principle provides an important point of difference between technology and science – it originates outside the body of scientific knowledge and comes into being to serve some innately technological purpose. The laws of physics may be used to analyze such things as air foils, propellers, and rivets once their operational principle has been devised, and they may even help in devising it; they in no way, however, contain or by themselves imply the principle. Polanyi makes essentially the same point a bit differently: "The physico-chemical topography of the object may in some cases serve as a clue to its technical interpretation, but by itself it would leave us completely in the dark [about how it achieves its operational purposes] . . . *The complete* [i.e., scientific] *knowledge of a machine as an object tells us nothing about it as a machine.*"

Apparently, both Vincenti and Polanyi believe that there is a gap between the functional (technological) and structural (scientific) description of technological objects. Nevertheless, engineers succeed in establishing more or less reliable connections between functional and structural descriptions of objects in technological explanations.

In order to analyze how engineers manage to bridge the gap between a structural and functional description of the world, we will now examine in more detail the structure of a technological explanation. As an example, we will use the explanation of the operation of one of the earliest types of steam engine, namely the Newcomen engine. The notion of operation refers to the way the function of this engine is realized. This type of steam engine was the first to be applied in practice for draining mines (from about 1712). Until the invention of the reciprocating steam engine at the end of the 18th century, Newcomen engines had primarily one function: They were used for *pumping water* (Hills, 1970, p. 134). Occasionally, steam engines were also used for other purposes, such as raising coal or ore from mines. During the 1770s, Smeaton, for instance, built several "winding engines" (Skempton, 1981, pp. 189–192). He also designed and built steam engines used for blowing furnaces. What is important, however, is that in these cases the steam engines did not drive the winding and blowing machinery *directly*, but indirectly through water wheels. The primary function of a steam engine was to pump water from a lower reservoir to a higher one, from which water flowed to a waterwheel. In other words, in these cases also the steam engines were driving pumps for raising water – "a business for which the fire-engine seems peculiarly adapted," according to Smeaton (Hills, 1970, p. 136). Thus, Newcomen engines were engines for driving pumps, and this was their main *technological function* until the end of the 18th century.

The operation of a Newcomen engine is based on a piston moving up and down in a cylinder (see Fig. 1). From a *technological* point of view, the operation of the Newcomen engine as a pumping device was well understood already early in the 18th century. The earliest descriptions of steam engines, for instance by Triewald (1734) and Desaguliers (1744), contain all the key elements for arguing from the input of steam engines, fuel in the form of coal or wood, to its output, the raising of water or the reciprocating motion of the great beam (which in practice was only used for driving pumps). The role of fire and steam in the technological operation of steam engines was clearly understood. In other words, the main technological function of steam engines could be explained in terms of the design of steam engines, as well as some relevant empirical facts and actions necessary for operating them. We will have a closer look at some of the essential ingredients of this technological explanation.

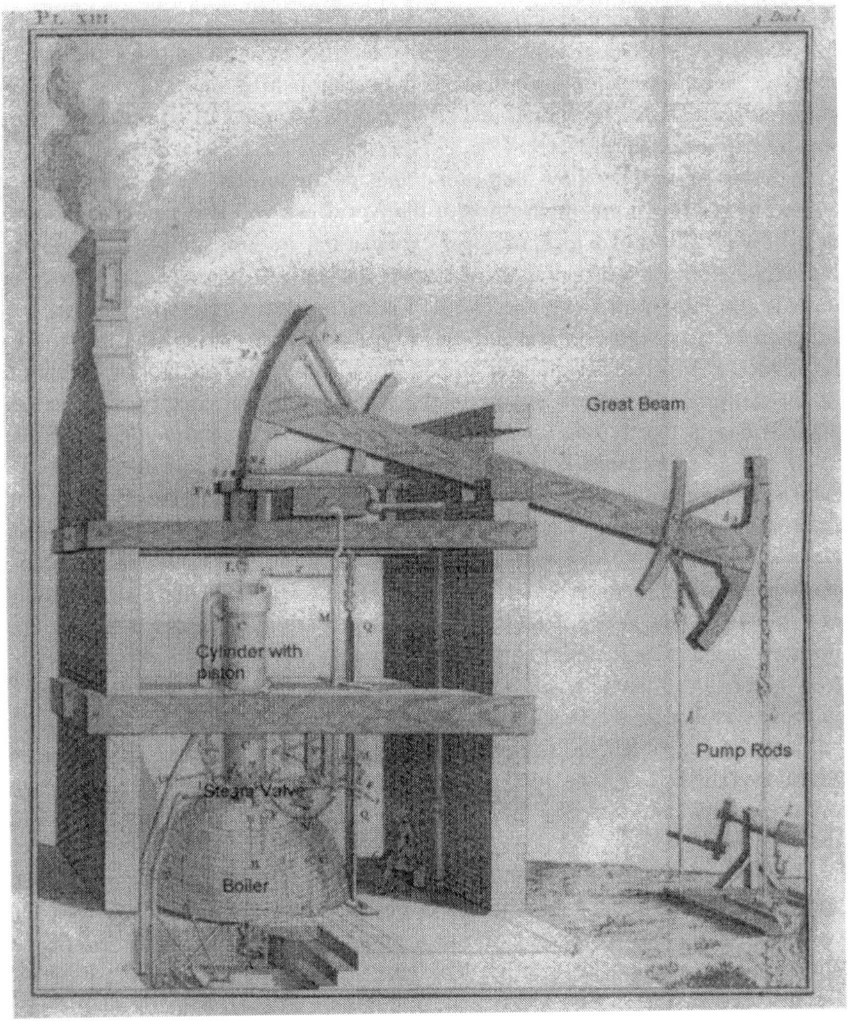

Figure 1.

First of all, it was a well-established empirical fact that fire had the "power" to transform water into steam, which occupies a much larger volume than water. According to Triewald, who considers steam to be "nothing but moist air", this fact is the basic principle underlying the operation of steam engines: "From this remarkable quality of the air of expanding when heated is derived the power or effect of the fire-machine".[9] Desaguliers also considers this to be the working principle of steam engines. Almost a century later, this is still considered by Rees to be the 'general principle of the steam-engine'.[10] The transformation of water into steam takes place in the boiler.

A second important step in explaining how fire, the "input" of steam engines, can bring about mechanical motion the "output" concerns the creation of a partial vacuum with the help of steam. This is achieved in the following way. When the steam valve is opened, steam flows into the cylinder. Because of the pressure of the steam and the weight of the water in the pump pipes, the piston will move upward until it reaches the top of the cylinder. At that moment, the steam valve is closed. Cold water is then injected into the cylinder, causing the condensation of steam and the creation a partial vacuum. The creation of a vacuum in a vessel is, furthermore, accompanied by the creation of a force, namely that of the atmosphere on the walls of the vessel. The existence of this force on an evacuated vessel had been amply demonstrated by the experiments of Otto von Guericke described in his *Experimenta Nova Magdeburgica* (1672, but reporting on experiments from the 1640s and 1650s). By means of a suitable *design*, this force can be exploited as a source of power for raising water. In the design of the Newcomen engine, one of the walls of the vessel is a movable piston; the creation of a vacuum causes this piston to move downward, because the condensing steam "is unable to withstand the weight of the atmosphere upon the piston, which is then hastily pressed down" (Triewald 1734, p. 12). Finally, the force of the atmosphere on this piston is transferred by a mechanical construction (the great beam) to the ends of pumping rods, which are lifted by the force of the atmosphere. As soon as the piston reaches the bottom of the cylinder, the steam valve reopens and the whole cycle is repeated.

The above chain of reasoning shows that the early steam-engine engineers had a detailed understanding of the way in which fire could generate power (for driving pumps) *in a Newcomen engine*. Their understanding was based, on the one hand, on some well-known physical phenomena, namely the expansive action of fire on water, the creation of a vacuum by the cooling (condensation) of steam, and the force of the atmosphere. On the other hand, the design of the Newcomen engine as well as certain actions (the opening and closing of valves) play an essential role. Without recourse to the detailed design and construction of a Newcomen engine and its mode of operation, they would not have been able to explain how fire could generate mechanical motion. Schematically, the explanation of the operation of Newcomen engines thus has the following form:

Scheme I

Explanans: – Description of physical phenomena
– Description of the structure (design) of the artifact
– Description of a series of actions

Explanandum: – Description of the function of the artifact

With regard to the issue of the relation between structure and function, the crucial question now is how explanans and explanandum are related to each other. In the case of Newcomen engines, the explanandum appears to follow logically from the explanans. The explanans describes in detail a causal mechanism that necessarily leads to the up and down motion of the pump rods. The function of the Newcomen engine – to drive pumps – seems therefore to be reduced to its structure. Note that the notion of structure has to be taken in a broad sense; it includes not only the design of the engine, but also the relevant physical phenomena and the actions necessary for operating the machine. So, contrary to earlier remarks, it looks as if in this example at least there is no gap between structure and function at all. This technological explanation appears to satisfy the conditions of a deductive-nomological explanation. A closer look reveals that this is not the case for two reasons. In more detail, the technological explanation of the function of Newcomen engines has the following form:

Scheme II

Explanans:

(1) physical phenomena:
- transformation of water into steam greatly increases its volume
- cooling of steam in a closed vessel creates a vacuum
- the atmosphere exerts a force of 1 kilogram per square centimeter
- etc.

(2) design of the engine:
- the steam engine consists of boiler, cylinder, piston, great beam, etc.
- the piston may move up and down in the cylinder
- the piston is connected to the great beam by a chain
- etc.

(3) a series of actions:
- opening of the steam valve causes steam to enter the cylinder and the piston to move upward
- closing of the steam valve and injection of cold water creates a vacuum in the cylinder
- etc.

Explanandum: Newcomen engines are a means to move the pump rods up and down, that is, to drive pumps (the function of steam engines).

The first reason why this is not an explanation of a function in terms of structure is that the explanans contains all kinds of functional concepts. Such notions as piston, cylinder, steam pipe, steam valve, etc. are all of a functional nature. Especially the descriptions of the design of the engine and of the actions for operating the engine are contaminated by functional concepts.

Can this contamination be undone? Is it possible to rephrase the explanans such that all functional concepts are eliminated and that nevertheless the explanation of the function is not endangered? In principle, this appears to be possible. The design of the Newcomen engine can be described by means of drawings and specifications for materials to be used, etc. For instance, a piston can be described in a functional way as an object intended to enclose a volume of steam in a cylinder in a moveable way. But it can also be described structurally by specifying its geometrical form and other physical properties. The same appears to apply to functional concepts used in describing the actions necessary for operating the engine; the opening of the steam valve, for instance, can be described as a change in the state of a subsystem of the engine. Let us make the assumption that, in principle, the explanans can be purified in this way of all functional concepts without affecting the validity of the explanation. Then, the chain of causally connected events described in the explanans with the help of functional concepts can be turned into a chain of causally connected events described with structural-physical concepts, such that the explanandum – the function of the engine – stays the same. (The explanation becomes much more complicated, but that only concerns the pragmatic aspects of this explanation.) In the light of this rather plausible assumption, the first argument against the claim that Scheme II is a technological explanation of a function in terms of structure becomes not very convincing.

There is still a second argument that this is not really an explanation of a function in terms of structure. It concerns the explanandum. According to Scheme I, the explanandum contains a description of the function of the Newcomen engine, i.e., to drive pumps, more in particular to move the pump rods up and down. Does that explanandum really follow from the explanans? The answer is negative. The explanans implies that *the great beam of the Newcomen engine will move the pump rods up and down, not that it is the function of this engine to move those pump rods up and down*. In other words, when we replace the explanandum in Scheme II by the statement that the pump rods move up and down, we end up with a paradigmatic case of a causal explanation of certain events.

The foregoing means that we have to distinguish carefully between the following two statements: (1) The Newcomen engine moves the pump rods up and down, and (2) the function of the Newcomen engine is to move the pump rods up and down. The first statement describes a factual cause-effect relation: The Newcomen engine produces a motion of the great beam, and this motion causes the pump rods to move. The explanans of Scheme II provides an acceptable explanation of this statement. The second statement is of a very different kind. It says that the Newcomen engine is a means for achieving a certain goal, namely the up and down motion of the pump rods. This statement does not follow from the explanans.[11] This may be seen as follows. Suppose that the motion of the great beam is used to drive a flywheel in order to produce a rotary motion of a shaft, which in turn drives, for instance, spinning machines. Then the function of the *same* object is the driving of spinning machines instead of the driving of pumps (or the driving of a construction that transforms a reciprocating motion into a rotary motion). In that situation, the Newcomen engine has a totally different function; it has become a means for the realization of a different end. Nevertheless, nothing has been changed in the explanans. So the conclusion would have to be that it would be possible to explain on the basis of one and the same explanans totally different functions in a logically deductive way.[12]

On the basis of the foregoing, we may conclude that the given explanation of the operation of the Newcomen engine is after all not a *technological* explanation, i.e., is not an explanation of function in terms of structure. So far, we have been unable to establish any link between the structural description of the Newcomen engine and the description of its function. Starting from the structure of the technological object, some physical principles, and a series of actions it is possible to causally explain a phenomenon with certain properties, i.e., the up and down motion of the great beam. But to go from there in a logically deductive way to the function of the object is not possible, because the same physical phenomenon can be, depending on the context, a means to quite different ends, that is, have different functions.

Given that the path from structure to function appears to run into a dead-end, let us see how far we can get by approaching the problem the other way around, namely from function to structure. On the basis of the assumption that the function of a Newcomen engine is to drive pumps of a certain type, it is possible to determine what kind of output the engine has to deliver. For driving the pumps, a specific type of up and down motion was required. That motion was necessary for the pumps to operate properly. Thus, from the structure of the pumps it was possible to draw up a list of specifications for the required output of the Newcomen engine. These specifications are described in a structural way. In every design process, this translation of a required function in structural properties of the system to be designed plays a crucial role. Let us assume that this translation has the character of a logical deduction.[13] In other words, when x has function F, $F(x)$, then this implies logically that x has certain structural properties (i.e., satisfies certain specifications) $S_1(x), \ldots, Sn(x)$:

$$F(x) \; -> \; S_1(x), \ldots, S_n(x).$$

Suppose now that system y has these properties (satisfies the specifications). May we then conclude that system y has function F? That conclusion is highly problematic, for the argument has the following form:

$$\frac{\begin{array}{c} F(x) -> S_1(x), \ldots, S_n(x) \\ S_1(y), \ldots, S_n(y) \end{array}}{F(y)}$$

This is a fallacy known as the affirmation of the consequence; a set of properties does not logically imply a function. Again, we reach the conclusion that on the basis of structural properties of an object, its function cannot be deduced in a logically deductive way.

The foregoing analysis of the relation between structure and function in the case of the Newcomen engine can be summarized in the following way:

Scheme III

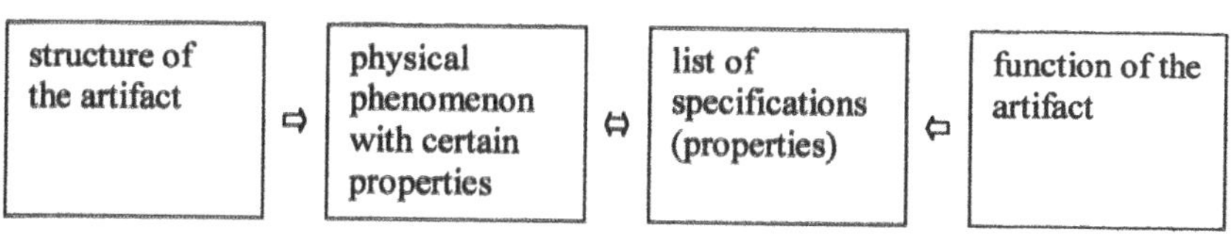

This scheme represents the complete chain of argument used in the technological explanation of the operation of this kind of steam engine.

It is a matter of fact that engineers succeed in establishing strong links between the structure and function of technological objects on the basis of technological explanations that have the structure represented in Scheme III. Since the two arrows in that scheme point in opposite directions, the relation between structure and function is not of a logical-deductive character. But what then is the nature of the link between structure and function?

Let us merely suggest an answer to this question. Suppose that a causal relation exists between two events *X* and *Y*:

(a) *X* is the cause of *Y*.

Whenever *X* takes place, *Y* will occur (under the usual *ceteris paribus* condition that no disturbances occur). Suppose, furthermore, that it is technologically possible to bring about the occurrence of event *X*. Then on the basis of the causal relation *a*, it is possible to draw up the following rule of action or pragmatic maxim:

(b) To realize *Y*, see to it that *X* takes place.

According to this rule of action, bringing about *X* is a means to let *Y* take place, in other words, bringing about *X* has the *function* of producing *Y*. Note that the rule of action *b* cannot be derived from the causal relation *a* in a logically deductive way. Nevertheless, we are dealing here with a (technological) rule of action that from a pragmatic point of view may prove to be highly reliable. Whether that is the case depends on whether the causal relation *a* is empirically well established, and whether in the given context of action the *ceteris paribus* clause holds.

Let us now apply this to the example of the Newcomen engine. For the *combined* system of the Newcomen engine and the pump, Scheme II contains a description of the causal chain of events that will lead to the up and down motion of the pump rods. When we refer, for the sake of brevity, to the explanans of Scheme II as the 'putting into operation of the Newcomen engine', then the following causal claim can be made:

(a') Putting into operation the Newcomen engine (*X*) will cause the up and down motion of the pump rods (*Y*).

On the basis of this claim, we formulate the following technological rule of action:

(b') To move the pump rods up and down (*Y*), put the Newcomen engine into operation (*X*).

This rule of action describes how a certain aim – the up and down motion of the pump rods – can be achieved by performing actions with a Newcomen engine. Within that context of action, the Newcomen engine becomes a means to an end, that is, acquires a function. That function is to drive pumps. The Newcomen engine, however, can perform that function only by virtue of its physical structure, since rule of action *b'* is based on causal relation *a'*, and the latter can only be derived on the basis of the detailed structure or design of the engine. According to this line of thought, the transformation of causal relations into pragmatic maxims (a transformation that does not have the form of a logical deduction) makes it possible to bridge the gap between structure and function in a technological design. A technological explanation, therefore, is not a deductive explanation, but it connects structure and function on the basis of causal relations and pragmatic rules of actions based on these causal relations.

The example of the Newcomen engine illustrates that the explanation of functions in terms of physical structures, something that is quite common in technology, raises interesting philosophical issues of a rather fundamental nature. The same example could also be used to illustrate epistemological and methodological issues in the philosophy of technological design.[14] For example, soon after the introduction of steam power technology, the problem of comparing the performance of different steam engines arose. In practice, this turned out to be an extremely complicated problem. To solve it, engineers developed a whole body of technological knowledge. They introduced such concepts as "foot-pounds per minute", "horsepower", and "duty". These concepts had no general physical significance, but were tailor made for specific types of engines in particular contexts of operation.

With the help of these concepts, engineers developed a techno-economic notion of efficiency. This notion of efficiency had a very restricted domain of application: It was useful only for a particular class of technological devices operating under very special conditions. In other words, the knowledge developed by engineers was *artifact-oriented* (design specific) and *context-dependent.* Only later on did some of these concepts acquire a more general meaning, before finally becoming incorporated into scientific theories, such as mechanics and thermodynamics. This development made it possible to apply these concepts to whatever kind of steam engine in whatever context of operation. Another problem faced by early steam engineers was how to translate vague customer demands with regard to steam engines into clear specifications for steam engines. The origin of the notion of "horsepower" is in this respect enlightening: It was quite common for customers to specify the required output of a steam engine in terms of the number of horses the engine was to replace. This kind of specification is not very useful for design engineers (what type of horses, working under what kind of conditions, and for how many hours a day, etc.?). The solution to this problem was of great importance for the development of steam power technology.[15] These examples show that technology itself, and not only the consequences of applying technology, may be a rich source for philosophical reflection; but they also show that philosophical reflection has to be based on detailed empirical case studies.

6. CONCLUDING REMARKS: THE EMPIRICAL TURN IN THE PHILOSOPHY OF TECHNOLOGY AND CRITICAL REFLECTION

This chapter started with the statement that the aim of philosophy of technology is a clarification, on the basis of a *critical* analysis, of concepts and conceptual frameworks concerning technology. In this final section, I will make some brief remarks on the role of empirical case studies in critical analysis.

Let me first say a few words about the possibility of critical analysis in general. Schematically, a critical analysis of concepts, conceptual frameworks, and the claims made within the context of those conceptual frameworks may take place at least from the following perspectives.

- *The logical-analytical perspective*. The main criteria for evaluating conceptual frameworks within this perspective are consistency, coherence, and the avoidance of equivocations. This leads to an immanent type of criticism (assuming that coherence is an intrinsic property of conceptual frameworks).
- *The empirical perspective*. This is the perspective of the empirical sciences. In these sciences, conceptual frameworks are evaluated primarily with regard to the empirical knowledge claims they make. Various criteria are used for evaluating knowledge claims, such as empirical adequacy, reliability, truth, explanatory power, etc.
- *The normative perspective*. In this perspective, conceptual frameworks are evaluated with regard to their normative implications. Various criteria are used to assess these normative implications, such as desirability, acceptability, justice, goodness, moral quality, etc.

Critical analysis of conceptual frameworks within philosophy is traditionally confined to the logical-analytical and normative perspective. Note that conceptual clarification is not an activity to which philosophy has an exclusive right; it is also an integral part of the empirical sciences.

My plea for an empirical turn in the philosophy of technology should not be misunderstood. It is not my intention to argue that the philosophy of technology should become a member of the empirical sciences and conduct its critical analysis mainly from the empirical perspective. In my opinion, questions like the one discussed above concerning the relation between structure and function in technological objects cannot be resolved in an empirical way: It is primarily a question about the *coherence* and *consistency* of two different ways of describing the world, or more in particular, two different ways of describing technological objects. This is a problem at such a high level of abstraction that it is not clear how a particular vision on the relation between structure and function of technological artifacts could be put to an empirical test, separately or as part of a more comprehensive conceptual framework concerning technology.

Nevertheless, the claim is put forward that empirical case studies are highly relevant – but not because the philosophy of technology is in need of an empirical anvil for testing its claims. On condition that these case studies provide empirically adequate descriptions of parts/aspects of technology, they contain the most interesting and relevant concepts and conceptual frameworks concerning technology. Instead of subjecting preconceived ideas about technology to a critical analysis, philosophers of technology should take as their starting point the concepts and conceptual frameworks that, implicitly or explicitly, lie at the basis of the best available empirical descriptions of the phenomenon called "technology". Those concepts and conceptual frameworks have survived the test of empirical adequacy, but that does not guarantee that they are free of conceptual problems and will survive a critical analysis from a logical-analytical perspective (think of quantum mechanics). Such an analysis may show incoherences or inconsistencies in those conceptual frameworks, which may necessitate a reformulation of basic concepts underlying the empirical descriptions of technological phenomena. What I have been arguing for here is, therefore, not an empirical philosophy of technology, but an empirically informed philosophy of technology.

NOTES

1. To illustrate this imbalance, the *Philosopher's Index* of Sept. 1997 contains 28.273 (6.095) entries about science (philosophy of science) and 2.765 (532) entries about technology (philosophy of technology).

2. For more details about the SPT, see Mitcham (1994), pp. 7-12.

3. See for instance the discussion between Joseph Pitt and Frederick Ferré on what philosophy of technology is about in Durbin (1990), pp. 3-23.

4. A simple glance at the content of the proceedings of SPT conferences illustrates the diversity of approaches and of topics addressed in the philosophy of technology; see *Philosophy and Technology*, vol. 1 (1983) – vol. 11 (1995). See also *Research in Philosophy and Technology* (Greenwich, CT: JAI Press), vol. 1 (1978) – vol. 13 (1993). For a brief survey of the field of the philosophy of technology, see P. A. Kroes (1998). See also C. Mitcham (1994).

5. For more on this point see Kroes (1997).

6. Note that the examples Pitt (1990) mentions illustrate the possible pragmatic relevance of a philosophy of technology focussing on these problems.

7. This section is a condensed version of Kroes (1998b).

8. The overall design process may be divided in various phases or steps, that correspond to distinguishable aspects of solving a design problem. Within design methodology the triad 'analysis-synthesis-evaluation' is often taken as a starting point for modeling the

design process. As long as designing is an activity performed by one single individual, these phases are relevant only from a conceptual point of view. As soon as designing becomes a matter of teamwork, which is by and large the situation in modern industry dealing with complex and large systems, the phasing of the design process becomes an important institutional tool for organizing, controlling and steering the process of product development. A typical example of the division of the design process in various phases, as part of a systems life cycle, is the following: requirements and specifications – preliminary conceptual design – logical design and architectural specs – detailed design and testing – operational implementation – evaluation and modification – operational deployment (for more details, see Sage, 1992). There are, of course, also interactions between various phases. As a model of the design process, this contains several intermediate steps between the starting point of a product development (the fixation of the commercial/functional requirements specification) and the end point of the design and development phase (the detailed technical design of the artifact). Gradually during the design process, the artifact takes on a definite shape until finally the whole artifact with all its components is uniquely determined in terms of its functional and physical properties.

9. Triewald (1734, p. 22). If steam is moist air, then the question arises from where all the air comes when water is transformed into steam. Triewald's answer is that (p. 23): "all water contains an unmeasurable quantity of air, which can easily be proved by allowing water to stand under the receiver of an air-pump."

10. See Rees (1819, p. 60). Rees in fact distinguishes between two main principles, (i) the generation of high pressure steam by heating water, which explains the operation of high pressure engines (see also p. 137), and (ii) the creation of a vacuum by condensing steam, which lies at the basis of the operation of the atmospheric machines (see also p. 75).

11. Compare the following example: the motion of the moon around the earth explains causally the phenomenon of ebb and flood; but the fact that this motion explains this phenomenon does not imply that it is the function of the moon-earth system to produce ebb and flood.

12. In this respect it is important to pay close attention to the question what is considered to be part of the Newcomen engine and what not. The pump rods or the fly-wheel construction are, contrary to the great beam, usually not taken to be part of the engine. This means that the "output" of this machine is the up and down motion of the great beam. This motion is causally explained by the explanans of Scheme II. When another system, for instance a pump or a fly-wheel construction, is coupled to this great beam, then it is possible, given enough knowledge about that system and the way it is coupled to the Newcomen engine, to causally explain the effect of the motion of the great beam on that system. But the causal explanation of the up and down motion of the great beam of the Newcomen engine does not imply an explanation of the fact that in one context the function of the engine is to drive pumps, in another to drive a fly-wheel construction.

13. We will not discuss the issue whether in reality this is the case; when the relation between function and specifications is of a weaker type than a logical deduction, the following analysis still holds.

14. For more details about the following, see Kroes (1991, pp. 79-97).

15. The translation of vague customer requirements in clear, preferably quantitative, specifications is an important problem in almost every field of technology (think of, e.g., user-friendly software, or sustainable products). Vincenti (1990, ch. 3) discusses a beautiful example, namely flying qualities, from the history of aeronautics.

REFERENCES

Borgmann, A. (1995). Does Philosophy Matter? *Technology in Society, 17*(3), 295–309.

Desaguliers. (1744). *A Course of Experimental Philosophy*. Vol. I (1734), vol. II (1744).

Durbin, P. T. (Ed.) (1990). *Broad and Narrow Interpretations of Philosophy of Technology. Philosophy of Technology*, vol. 7 Dordrecht: Kluwer.

Hills, R. I. (1970). *Power in the Industrial Revolution*. Manchester: Manchester University Press.

Ihde, D. (1993). *Philosophy of Technology: An Introduction*. New York: Paragon House.

Kroes, P. A. (1991). Steam Engines and the Concept of Efficiency: Characteristics of Technological Knowledge. *Methodology and Science, 24*(2), 79–97.

Kroes, P. A. (1997). Carl Mitcham: Thinking through Technology: The Path between Engineering and Philosophy, review in *ANTW, 89*(1), 103–104.

Kroes, P. A. (1998a). Philosophy of Technology. In: E. Graig (Ed.), *Routledge Encyclopedia of Philosophy*, vol. 9 (pp. 284–288). London: Routledge.

Kroes, P. A. (1998b). Technological Explanations: The Relation between Structure and Function of Technological Objects. *Technè, 3*(3), 11. (Electronic Journal of the Society for Philosophy and Technology http://scholar.lib.vt.edu/ejournals/SPT/v3n3/html/KROES.html.)

Mitcham, C. (1994). *Thinking through Technology: The Path between Engineering and Philosophy*. Chicago: Univ. of Chicago Press.

Pitt, J. (1990). In Search of a New Prometheus. In: P. T. Durbin (Ed.).

Rees, A. (1819). *The Cyclopedia, or Universal Dictionary of Arts, Sciences, and Literature*. London.

Sage, A. P. (1992). *Systems Engineering*. New York: John Wiley & Sons.

Searle, J. R. (1995). *The Construction of Social Reality*. New York: Penguin Books.

Skempton, A. W. (1981). *John Smeaton, FRS*. London: Thomas Telford, pp. 189-192.

Ströker, E. (1938). Philosophy of Technology: Problems of a Philosophical Discipline. In: P. T. Durbin & F. Rapp (Eds.), *Philosophy and Technology* (pp. 323–336). Dordrecht: Reidel.

Triewald, M. (1928)[1734]. *Short Description of the Atmospheric Engine*. Cambridge: W. Heffer & Sons.

Vincenti, W. G. (1990). *What Engineers Know and How They Know It*. Baltimore: John Hopkins University Press.

TO WHAT EXTENT SHOULD A CRITICAL PHILOSOPHY OF TECHNOLOGY BE CONSTRUCTIVIST?

Bram Bos

Should there be an empirical turn in the philosophy of technology? The desire to give more empirical backing to an overly theoretical field echoes a development that took place in science studies almost thirty years ago. Back in those days, a group of young turks within the sociology of science wanted to get rid of the highly theoretical way philosophers of science tried to explain how science attained or approximated truth, or justified knowledge. Previously, sociologists restricted themselves professionally to the "social dimensions" of science, evading any sociological claim about the actual content, truthfulness, or rationality of the disciplines under study. With the Edinburgh school and its successors the actual content of scientific claims became open to sociological clarification. This "sociological turn" meant a challenge to the philosophy of science, because it cast doubt on some of its most cherished beliefs. During the following three decades a broad area of empirical investigation emerged under the common heading of Science Studies.

The fact that philosophers of technology now call for an empirical turn in this specific area of philosophical investigation, seems to confirm at first glance the prejudice that the philosophy of technology is the lesser and in this case especially slower relative of the philosophy of science. More importantly, it shows how disparate both these fields of philosophical inquiry have been, while their respective subject matters are commonly seen to be quite closely connected, if not derived from one another. Moreover, Science Studies has

The Empirical Turn in the Philosophy of Technology, Volume 20, pages 45–64.

ISBN: 0-7623-0755-2

evolved into Science and Technology Studies (STS) in the past twenty years, since the same sociological and anthropological methods turned out to be fruitful not only for describing the way science produces knowledge, but also the way technical artifacts and systems enter the natural and social world. For some, this relation between science and technology is worked out mainly methodologically – as for instance with Wiebe Bijker, who translated the Empirical Programme of Relativism (EPOR) at Bath University for studying science into a research program for describing the social construction processes from which technical artifacts originate. Others identify far stronger links between science and technology and propose an ontology in which the development and survival of knowledge and technical objects can be studied in essentially the same way, since in both cases their existence is due to stable networks of relationships (Bruno Latour, Michel Callon).

Ever since this specific branch of technology studies within STS emerged, proponents have claimed to offer a better view on technology and technical change than the preceding philosophy of technology or technology assessment studies. The prime reason for this claim has been their emphasis on detailed empirical case studies, in contrast with then current philosophical theories of technology, which seemed to use real world examples only to explain their universal and abstract categories and mechanisms. According to people such as Bijker and Latour these case studies show the contingent character of realized technical artifacts. Technical artifacts come into existence by construction-processes, characterized by the struggle between a diversity of social groups over the meaning of the artifact under construction. The key is that no privileged factor or actor alone determines in advance the outcome of these processes. The end result cannot be explained satisfactorily by reference to, for instance, a technocratic class of engineers, a universally valid type of technical rationality, or a set of necessary technical laws alone. Hence, the fundamental idea within many classical studies in the philosophy of technology, that technology evolves according to its own logic or its own rationality, was considered to have been proved false by the constructivists.

Thus, from the start technology studies has opposed itself to the philosophy of technology, and justified this primarily by referring to the empirical character of their research, thereby uncovering what technology and technical change "really" is. We should keep in mind however, that constructivism aimed at a specific – but prominent – part of the philosophy of technology, namely those philosophies that portray technology as forming an ontological class in itself, fundamentally distinct from nature and humans, and guided by its own principles, that are strange to the way nature or social relations are organized. Typical representatives of this strand are Martin Heidegger and Jacques Ellul.

This opposition between constructivism and (a specific part of) the philosophy of technology is significant enough to work out in greater detail, since it highlights the two extremes which scholars who want to develop a normative or critical theory of technology have to navigate between. In this article I will focus on a specific part of the philosophy of technology, namely that part which aims at a critical evaluation of technology – or rather, specific technologies – in the light of its societal embeddedness. The central question will be, whether this particular strand of philosophy of technology would benefit from an empirical turn", and – subsequently – whether such an empirical turn should be constructivist or not.

I will answer the first question in the affirmative. A philosophy of technology which strives for a critical analysis and evaluation of the role that technologies play in our daily life and society, should indeed have empirical research as a necessary constituent. The main argument for this is that technologies acquire their normative or political relevance only within their specific context. Both technical artifacts and technical systems function only within a specific environment that is social, technical, economical, and bureaucratic all at once. The actual normative and political features of such sociotechnical ensembles" reside in their specific constellation, in which neither the technical, nor the social or the economical can be said to be a privileged determinant beforehand. It should be noted that this argument entails a specific conceptualization of technology, which is akin to that used in constructivism. This is not to say – and this pertains to the second question – that a critical philosophical theory of technology should be constructivist in all respects. On the contrary: I will argue that constructivism cannot be sufficient either as a theoretical framework, or as an empirical methodology for a philosophy of technology that wants to analyze existing and developing sociotechnical systems and criticize their constitution as well.

1. CONSTRUCTIVISM: WHAT BOX IS ACTUALLY OPENED?

Constructivist studies in technology are characterized by a variety of different approaches, which however do have in common some general features. One of the most basic tenets constructivists share is that new technical artifacts are the result of historically *contingent* processes, and thus are in no way determined by universal, time-independent factors. A second tenet is the fundamental flexibility of meanings attributed to an artifact-under-construction. The "same" object can be interpreted in very different ways by different actors.

A well-known example is Bijker's study of the social construction of the bicycle (Bijker, 1995). Back in the days of the first bicycles, a rather wide array of different types existed. Bijker shows how different social groups (young men, housewives, the elderly) preferred different bikes for different reasons. Young men liked high-wheeled *Ordinaries,* just because of the fact that it was a risky affair to ride such a bike with its high front wheel and very tiny rear wheel. Housewives and elderly people preferred the *Safety Ordinary* – equipped with a larger rear wheel than its predecessor – especially after rubber tires were added. Within a period of time, the array of different types of bikes was reduced to one standard type: the safety bike. Thus, the definition of what a bike was (its meaning) was not fixed from the start, but got its stability during a timely process of "negotiations" between different social groups. Bijker thus shows the *socially* constructed character of technical artifacts by means of historical-empirical analyses of their development.

Bijker and Trevor Pinch (1987) explicitly interpret this result as an empirical refutation of the quite dominant view in the philosophy of technology that technology possessed an autonomous character, or even has a determining role in shaping society, due to its fundamentally different and inevitable logic. Bijker argues that this view was a mistake: not only do social processes influence technical innovation processes, these processes can also be shown to be contingent. As one of their favorite mottoes states: "It could have been otherwise". The mistake can be attributed to the black box character of realized artifacts. Once negotiations on the meaning of an artifact are closed (which is a contingent matter), the resulting product seems to be the necessary (logical) outcome, since our interpretation of the meaning (its function, its role in our lives or society) is perfectly consonant with the design and structure of the artifact. After negotiations have ended, and controversy closed, the general pattern is that we forget the historical origins of an artifact, which makes us think the result was a necessary outcome of transcendent or universal technical criteria. The bike we know seems to be the *best* bike. Like a miracle, the artifact fits precisely with our definition of what it should be! We do not question its particular construction anymore, but treat it as a black box, an object of which only the inputs and outputs in actual usage are relevant for us.

Constructivism saw as its principal task to open this black box, by showing its constructed character and consequently its human and social origins. This was seen as a liberating move: We do not have to accept the idea that technology evolves along its own logic. Any technical artifact is the result of our own hands, and its construction history is not explained fully by referring to inescapable technical principles like efficiency, control, or blunt causality. Numerous choices have been made during its construction, choices which were not inevitable for those involved. These choices do not involve technical issues alone, but social, aesthetic, and normative issues as well. By opening the black box constructivist methodology tries to highlight all these different kinds of decisions. For *methodological* reasons, no one of these is given a privileged place a priori as a determining factor. Therefore, constructivism implies at least a methodologically relativist position.

The use of this approach is twofold. First, studies of concrete cases highlight the social and cultural values that codetermine that specific end result. Second, there is some mitigation of the general paralysis that technological determination tends to cause concerning societal possibilities for changing technology. Although constructivism does not imply that technological change is completely under our control, it does imply that all kinds of human action leave their mark upon technical artifacts. But since the end-result is the product of an aggregate of different actions of different groups, there is no specific group or specific point of action to be identified as the prime mover. There still is no steering wheel out there.

The type of constructivism sketched here is based primarily on the work of Bijker and Pinch (Bijker, Hughes & Pinch, 1987). Their Social Construction of Technology (SCOT) model is strongly cognitively oriented, which will not surprise us once we consider its origin in the sociology of science, whose basic conviction is that nature, or the materiality of things, in no way constrains the type of knowledge we cherish about the world (see for example Barnes, Bloor & Henry, 1996). Early SCOT analyses typically portray technical change as negotiations about meaning, definition and interpretation. Its methodological relativism thus parallels the epistemological relativism of the sociology of scientific knowledge.

This position, however, does not apply to constructivist technology studies as a whole. The Actor Network theory of Callon and Latour has proposed a more far-reaching variant of constructivist thought, in which a peculiar ontology plays a significant role. While SCOT can be said to be agnostic about the role and influence of the material world, of things and objects, AN theory integrates this role in an actor-oriented model of technical change. This should be taken quite literally: objects, technical artifacts, organisms and the like enter the stage as actors ("actants") in their own right. Construction processes are characterized by a continuous struggle between actors to acquire control over their respective actions. In order to successfully build a new technical artifact, an engineer has to play at different grounds at the same time. Financiers, the government, potential customers, apparatuses, atoms – they all have to be *enrolled* successfully in a stable network in which all actants act in concert, while still striving for their "own" interest. In order to successfully enroll actants in a network, a network-builder thus has to align all the different interests of the elements (actants) in such a network.

This idea of a network not only points to the connectedness between actants, but is also essential for explaining why technical artifacts function in a reliable way in different places under different circumstances. According to AN theory this reliability is not due to the intrinsic technical properties of the artifact, but to its being part of a stable network, consisting of different (economical, technical, social and normative) relations. For instance, an ordinary consumer car is not of much use in a desert without any fuel supplies, maintenance facilities, or roads. The car can only be perceived as a black box within this much broader network. Of course, we could make a different car for the desert, such as a four-wheel drive with special equipment on board for repairs, extended fuel supply, and the like. But even in this case, we need a remote network of fuel suppliers, factories to make spare parts, and we should be taught how to repair the car in case of breakdown. Thus, we can only permit ourselves not to bother about the inner workings of the artifact once a stable network in one way or the other exists around it.

Although the previous description is necessarily sketchy, I hope to have made clear that there are some noteworthy differences between the constructivist approaches of Bijker and AN theory. While Bijker escapes from the dichotomy between technology and society by taking up a methodologically relativist position, focusing almost exclusively on the cognitive and symbolic aspects of technical change, AN theory redefines the ontological framework in which technology and technological change is perceived. Instead of relegating material and causal factors to a domain that we can only know via their cognitive representation or interpretation (Bijker), Latour and Callon draw these material and causal factors into the sphere of action and interests, thereby placing them on the same ontological level as humans. Since AN theory denies a fundamental distinction between humans and non-humans, will and intentions may as well be attributed to non-humans in AN theoretical analyses.

Compared to Bijker's SCOT approach, AN theory is far more ambitious with respect to its domain of application. SCOT focuses almost exclusively on technical change at the level of artifacts. Once controversies are closed, there is not much substantial to be said from a SCOT perspective. By contrast, AN theory does not stop at the threshold of closure. Although most AN studies are directed towards the explanation of the eventual success or failure of specific network builders, the theoretical apparatus of AN theory does allow for useful descriptions and analyses of concrete existing technological systems.

This fruit comes however at considerable cost – a cost not everyone is willing to pay. For instance, many people would like to maintain the distinction between human and non-human actors. Especially if we want to deal with issues of normativity, intentionality and responsibility in technology and technical change, such a distinction seems to be of a certain relevance, at least to me.

2. THE EMPTY BLACK BOX: WINNER'S CRITIQUE OF CONSTRUCTIVISM

The aim of this article is to assess the relevance of constructivism for the philosophy of technology. As we have seen, constructivism has challenged from the start certain cherished ideas in "classical" philosophy of technology, especially that inspired by Heidegger. One of the members of this tradition is Langdon Winner, whose Autonomous Technology (1977) bears the influence of Jacques Ellul. Although his later The Whale and the Reactor (1986) is much more tentative about the possibilities for a rigorous political philosophical critique of technology, this continues to be his primary goal.

Winner's central idea is that contemporary technology represents a law-making force in its own right. By creating large technical systems such as nuclear energy plants, we not only introduce a means for producing certain ends in our society, the means themselves imply a certain political order. Nuclear plants possess a necessarily centralized character: the power over power is in the hands of a few people. Solar energy systems, by contrast, have the possibility of producing energy in a more decentralized manner. This specific technology leaves open the possibility that any household, any third-world peasant, might produce his or her own energy without being dependent on large (multi)national companies. Although it is quite conceivable that a large energy firm such as Shell would centralize solar energy as well – for instance, by building huge acres of solar panels in the Sahara – the solar energy technology itself does not imply this distribution of control. Thus, according to Winner, specific technologies may possess an inherent political character: "artifacts have politics." This may be because the adoption of a given technical system actually requires the creation and maintenance of a particular set of social conditions as the operating environment of that system (for instance nuclear energy), or – in a weaker sense – because a given kind of technology such as solar energy is strongly compatible with, but does not strictly require, social and political relationships of a particular kind (Winner, 1986, p. 32).

According to Winner, the highlighting *and* criticizing of political and normative features of technology or technological systems is a prime task for the philosophy of technology. How could constructivism contribute to this? If constructivism could live up to its own ambition to replace large parts of the philosophy of technology, this would be a legitimate question. Winner himself was not quite positive about this in a polemical contribution to the Biennial Conference of the Society for Philosophy and Technology in 1991, which appeared in 1993 in both *Science as Culture* and *Science, Technology and Human Values* (two journals oriented toward the STS community). He criticized the emerging trend of (social) constructivism at that time for a number of reasons, of which the most important was the gap between the pretences of constructivism, on the one hand, and their actual achievements, on the other. Constructivists opened the black box, but found it empty. At least, they found nothing of interest to a philosopher of technology such as Winner. "The most obvious lack in social constructivist writing is an almost total disregard for the social consequences of technical choice" (1993, p. 439). Winner explained this as a result of STS with those never-ending questions about the societal and environmental consequences of technology that old-style philosophy of technology. Constructivists focused on technological change (innovation) because it was new, and was more fun. As Winner points out, this enthusiasm was also born from the results the empirical turn in science studies had achieved, on which constructivist studies of technology were modeled in the first place.

Due to this focus on the empirical study of developmental processes in technology, constructivism has little to say about the ways technologies actually change our individual lives, our personal experiences and social relations. The end product is taken as a given, the origin of which is all that should be explained. This would not be so troublesome if constructivism tempered its pretensions, but it does not. Repeatedly, constructivists have argued that empirical investigations in the constructivist vein highlight the way technology "really" is. To show how technology is constructed is to show what technology is.

This equation of technology with technological change is indeed typical of constructivist studies. In STS literature it is quite common to use technology and technological change alternately. And this is an important source of misunderstanding between Winner and the constructivists he criticizes. I can illustrate this with the commonly used idea of opening the black box. In constructivist readings, this idea applies to the empirical and theoretical analysis of the historical processes that precede the existing consensus about definition and design of a specific technical artifact or system. If Bijker opens the black box of the bicycle, he does not need wrenches or screwdrivers. What he opens up is the taken-for-granted definition of what a bicycle is, by showing how this taken-for-granted-ness is the contingent result of a struggle between quite different definitions attached to it during its process of construction.

Although Winner himself is also somewhat ambiguous about his quest, he seems to prefer the more common sense, or literal notion of opening black boxes: get your toolbox in order to screw open the artifact under study. How does it work? In what way does this particular artifact exert political or normative force upon its surroundings? Where are the politics inside this artifact (Winner, 1980)? No wonder Winner finds the tools constructivism proposes of little use for his philosophy of technology. Part of his critique of constructivism thus originates from a different interpretation of what should be deconstructed in the process of analysis. Constructivism seems to be satisfied in deconstructing the self-evident character of technology, by showing the contingent origin of the definition of specific technical artifacts. What is highlighted is the process, not the product. By contrast, Winner is primarily concerned with the product and its place in our social context. A process-oriented analysis of the origin of the product does not supply us with criteria to formulate a substantial critique of its inherently political character, its value-ladenness.

3. A CONSTRUCTIVIST'S RESPONSE

Insofar as constructivists claim to offer the better alternative for the philosophy of technology in general, I think Winner's critique is correct. And, as time has passed, constructivists have become more modest in their opposition to approaches, almost in parallel with their becoming an established scientific field. One of them, Bijker, quite adequately responded to critics such as Winner in 1993 ("Do Not Despair: There is Life after Constructivism"), where he counters the argument that social constructivism leaves us without any means for criticizing and steering technological change, because of its inherent relativism. For him, early studies in constructivism offered a necessary antidote against technological-determinist thinking in society. It was meant to lift the hearts of all those out there who were caught in the black-painted thoughts of pessimists such as Ellul or Ivan Illich, who taught that technology was our inescapable and uncontrollable destiny. Constructivism was only meant as a "detour", in order to reach the "main route" again, after we have left the doom scenarios behind. Constructivism shows how people in fact have influenced technological development, and negates the idea that this development eventually has been governed by universal principles. So cheer up: We can do something about it! The relativism social constructivism entails is in no way a threat to this, since it only concerns methodological relativism. This does not imply either normative relativism, or ontological relativism. Social constructivism does not block the road to critical assessments or substantial critique. As Bijker put it: "on the basis of relativistic studies of technology, it is possible to argue for political and ethical positions with respect to technological choices" (1993, p. 116).

Thus, although Bijker agrees with Winner that constructivism up to that time did not contribute much to issues concerning the consequences or ethical implications of technology, there is no fundamental reason why this should not be possible. Indeed, Bijker actually proclaims in the same article that the time has come to leave the academic detour of empirical constructivist case studies, in order to deal with the practical issues in society that we are concerned with. Now that technological determinism is effectively debunked, and new ways of conceptualizing technology have been developed in constructivist theorizing, we should be able to answer the same questions from a different perspective.

These "same" questions concern – according to Bijker – the issue of control of and intervention in technological change. The old answers were directed toward either mitigating the negative side effects of otherwise autonomously technological developments or based on cost-benefit analyses. Bijker mentions the U.S. Office of Technology Assessment as the exemplary institution practicing both. Clearly these types of answers were heavily inspired by a technological determinist way of thinking. Once started, technological developments follow their own course. The two moments for intervention and control are at the very beginning (cost-benefit analyses) or at the end (try to avoid the negative side effects). Both these attempts for intervention face serious difficulties. The first one because a realistic cost-benefit analysis cannot be made at the very start, since the costs and the benefits only become clear along the way; the second because we often can only avoid the negative side effects of new technologies at great cost, since they already have become entrenched too deeply into society. Classical answers of control and intervention thus face the so-called Collingridge dilemma: Either it is too early to decide what intervention is actually needed or it is too late for intervention (Collingridge, 1980).

For Bijker this dilemma disappears if we adopt the constructivist perspective on technology, in which society and technology are inextricably intertwined into *sociotechnical ensembles*. If we accept this idea, "a multitude of opportunities to influence the development of technology (and society, and . . .)" (Bijker, 1993, p. 135) can be indicated. This almost sounds like a miracle. Just start thinking in a different way, and old problems disappear. It turns out not to be that simple. Bijker's subsequent suggestions about where to look for themes for studies "under the banner of control and intervention studies" turn out to be different answers to *different* questions. The key contribution of social constructivism with respect to practical problems seems to be twofold. First, the fact that social constructivist studies show the involvement of an array of different social groups in construction processes is used as a basis for voicing a plea for better involvement or participation of "social groups" in these processes. Patients, for example, should be more strongly involved in the shaping of clinical technologies (non-traditional design constituencies). This idea Bijker has worked out in somewhat more detail in his inaugural lecture of 1995 under the title "Democratizing Technological Culture". Second, social constructivism should extend its deconstructive capacities, in order to break up the self-evidence of norms and regulations in technological design processes, or the self-evident character of economic rationalizations concerning investments, efficiency, and cost-benefit relations.

Although these suggestions are rather worthwhile when one is dealing with concrete choices in technological innovation, it should be noted that they are in no way satisfying answers for the traditional questions of control and intervention. Bluntly put: while old-style technology assessment would ask how we can steer technological development, Bijker redefines the question into "who should be involved in technological development, and how?" And while old-style philosophers of technology will ask "what is right or wrong about (this specific piece of) technology", Bijker's answer seems to be: "Let the people decide".

Adopting a constructivist stance thus does not offer us new answers to old pressing questions, but new answers to other – equally valid – questions. With respect to the issue of control and intervention, social constructivism proposes *procedural* improvements in order to increase participation in design processes. The presupposition behind this is that more participation results in better technology. What "better" is cannot be decided, however, on the basis of social constructivist analyses. Neither can we decide on the basis of social constructivism where and how to intervene. Therefore, I think Bijker is wrong to claim that a social constructivist perspective in which society and technology are seen as inextricably interwoven (the "seamless web") really frees us from the control dilemma of Collingridge. On the contrary, all those active participants in social construction processes face the same dilemma as well. The only difference is that it is no longer a small elite of engineers, politicians, and technology assessment specialists who have to bear the burden.

The other contribution social constructivism might make outside the academy follows from its deconstructive capacities. Because no part of technology is in principle exempted from deconstructive analysis, we can assess actual technical artifacts and technological systems not only to show their constructed character, but also to investigate how specific design characteristics, technical norms, and dominant definitions of what the artifact or the system are about relate to specific groups, ways of thinking or even ideology. A political or normative critique of specific technological systems or sociotechnical ensembles thus will benefit from a constructivist analysis, as will be shown below when I discuss Andrew Feenberg's attempts to integrate constructivist thinking in a critical philosophical theory of technology.

Before that, I'd like to make some preliminary concluding remarks about the use of constructivist methods for the philosophy of technology. First, the important step constructivism made is the choice to conceive of technology as essentially interwoven with the social sphere. This implies, on the one hand, that in the description and explanation of the genesis of new technical artifacts and systems, entities, and actions of quite different (heterogeneous) origin play a role, without an a priori hierarchy between these entities or actions. Philosophies of technology might profit from this perspective, since it helps us to avoid the risk of interpreting the path of history with hindsight as a necessary and completely technically determined trajectory. Especially a critical philosophy of technology may, with this methodology, find clues relevant for critical assessment, which would otherwise remain hidden behind the facade of a technology defining its own conditions of development.

On the other hand, an extension of this methodological relativism to ontology for example in the way Latour does – would result in a conception of technology – that is fundamentally different from classical ways of portraying technology in the philosophy of technology. Instead of ascribing essential distinguishing ontological features to technology, portraying it as an analytically distinguishable phenomenon, or examining it as a differentiated subsystem in society, constructivism presents us with a picture of technology that cannot be understood otherwise than as a (variable) mixture of material elements, organisms like humans, and social institutions. The "essence" of technology is its heterogeneous constitution. Although I definitely do not share the ontological monism Latour promotes, I do think it is fruitful for a critical philosophy of technology to lay aside the ambition of defining technology in some universally applicable terms before actually assessing concrete instances of technology, artifacts as well as systems. Isolating the technical part of technological systems and artifacts a priori will significantly hamper the analysis of how social values, political goals, and meaning relate to their material concretization.

Furthermore, constructivist methodology is useful due to its deconstructive capacities. This may highlight features of technology that remain implicit if we treat technology as a phenomenon in its own right. This is especially relevant if we are interested in such things as the intrinsic normativity of technical regulations (Winner's politics of artifacts). Andrew Feenberg introduces the term "technical codes" for this.

However, constructivism, especially its social constructivist variant, deals with technical change. The actual substance of realized artifacts, the political order or normative guidelines they enforce, is of less concern. Politics of technology is interpreted without much ado as the politics of technological change. This is perfectly legitimate as long as it is recognized only as a useful or provisional limitation of the subject matter. For the purposes of a philosopher such as Winner, however, this is clearly not enough. Winner wants to identify the politics in artifacts (or technological systems), both emerging as well as existing ones. Empirical knowledge of the origin of these artifacts and systems will likely give us *some* insight into these political features, but we cannot claim to have a complete description of an entity once we have described its biography.

Take, for example, one of Winner's own favorite cases of an artifact that intrinsically has politics: the New York bridges on the Long Island Expressway. The city planner at that time, Robert Moses, deliberately constructed these bridges so low that buses would not be able to pass below them, while cars would. Since richer white people tended to have cars, while poorer blacks did not at that time (and thus had to take the bus), the political effect of these bridges was a separation of social class and racial groups. Whites could go to a place such as Jones Beach on Long Island, while black people could not.

Now if we make a social constructivist analysis of these bridges, we would likely come up with a list of the groups that were involved in their design: Moses, the local government, engineers – the list might not be that long. Subsequently we would make an inventory of the specific and varying ideas of what these bridges were meant for, how they should be interpreted. If Winner is right, the definition of these bridges (their perceived function) might be different for Moses compared to, for example, the communal government. At least, they bore a second function in Moses's interpretation: namely the function of a *social selection environment.* Finally, we could reconstruct how these different interpretations eventually converged, or were sorted out into closure, when the final decision with respect to the specific design was made.

Although this might not be the most interesting case study constructivism ever performed, it serves to illustrate some of the limitations of this kind of the social constructivist program – limitations Winner wants to transcend. First and foremost, if we did not know with hindsight that these bridges did have this selective effect between cars and buses, we would never count "people without cars" as a relevant social group in our list of actors. Just as we would not count people intending to commit suicide as a relevant social group, even though their definition of the function of a bridge might very well be rather different: a definition Mose's design of low bridges would likely not fit. Only with hindsight would we think of this category as a relevant social group, illegitimately excluded from the construction process. On the one hand, this shows as legitimate Bijker's plea for extending the range of participants in the design process, but we will run a good chance of not including everyone for whom the artifact turns out to be relevant!

Secondly, the intrinsic political effect of these bridges cannot be assessed by a constructivist analysis of its construction process. After all, the fact that specific social groups were not involved in its design does not imply they will necessarily suffer from its effects.

Moreover, the selective function of these bridges may well change over time, without the artifacts themselves being changed one molecule. For example, since some rich people in New York tend to have robust four wheel drives, too high to pass below the bridges, while at the same time even poorer city dwellers may have access to cars these days, the structural selective effect of these bridges has changed dramatically (I owe this observation to Rein de Wilde). This shows that artifacts have particular politics only within a particular historical context. Thus, contrary to what the term "intrinsic" suggests, there is nothing political in the isolated material object itself. However, this cannot be taken as an argument for the intrinsic neutrality of technical things, since all artifacts always exist within specific contexts.

Summarizing: when Winner polemically argues that constructivism opens the black box and finds it empty, he is right only in certain respects. If we expect constructivism to be a complete alternative for reflective and critical philosophies of technology, we expect too much. Part of this confusion is due to the fact that constructivism almost exclusively focuses on processes of change, while suggesting that this methodology will at the same time give us insight into what technology really is. Although we cannot deny a close relationship between the genesis of technological designs, and their actual constitution, we had better not equate them. At least, this seems a wise strategy for philosophers of technology interested in the actual political and normative roles technology plays in specific contexts.

4. FEENBERG'S ADD-ON TO CONSTRUCTIVISM

One good example of a philosopher trying to reap the fruits of constructivism while at the same time formulating a substantive critique of technology is Feenberg. He is among the few philosophical writers of the last two decades trying to transcend the boundaries between modern philosophical theorizing about technology and the empirical studies of technology originating in (social) constructivism and actor-network theory. His last book *Questioning Technology* (1999), deals explicitly with this issue.

Feenberg's work originally derives from Frankfurt School critical theory, especially the work of Herbert Marcuse. Feenberg is especially indebted to Marcuse's critique of Weber's concept of value-free "formal rationality". The rationality of the capitalist market, as well as the rationality of large-scale organizations, such as bureaucracies is formal in the sense that the rules and criteria of which these rationalities consist do not discriminate between values, groups, or people. This means, in Weberian terms, that they are value-free. For instance, everyone is treated equally in the capitalist market – as long as one brings one's money along.

We know from experience, however, as well as from Matthew 25:29 and Marxist critiques of capitalist economies, that the free market *does* result in inequalities, despite its formal value-free character. In fact, capitalist economies tend to preserve and even increase existing inequalities. This is called the formal bias of capitalist economies. No substantive difference is made between groups, but in practice there is a systematic bias favoring certain groups above others.

Likewise, Marcuse criticizes the idea of a value-free rationality in technology. As Feenberg puts it:

> Like market rationality, 'technological rationality' constitutes the basis for elite control of society. That control is not simply an extrinsic purpose served by neutral systems and machines but is internal to their very structure (Feenberg, 1991, p. 69).

The strong claim of Marcuse, which Feenberg adopts in his own theory, is that technological systems and machines have a bias in favor of dominant (elite) groups in society, which is internal to their rational structure. Characteristics of these systems and machines; that seem to be the necessary result of objective technical (or natural) constraints turn out to be socially determined, and subsequently determining of the social in a specific way. This does not mean they are not rational at all. The fundamental point of Marcuse is that technical rationality sustains power relations, although he does not dispute the validity of this rationality itself. The critical task of the analyst is to uncover these power-knowledge relationships.

One of the favorite examples in Feenberg's earlier work is the structuring of work-relations in factories. One of the key notions in management theory and practice is the concept of efficiency, where this value is often taken to be neutral or value-free. One could define it formally as the ratio of inputs to outputs (1999, p. 160). However, in order to apply such a formal notion, we need to know what counts as input or output, whether we can acquire them or not, what counts as waste instead of output, etc. These are highly socially determined considerations, and therefore, efficiency is as well. Within a factory setting, efficiency is often defined against the " background assumption" of worker resistance to work (Feenberg, 1991, p. 69). In order to counter this resistant attitude, factory machines are designed to de-skill and discipline workers (Braverman, 1974). The result is known as Fordism or Taylorism. Efficiency, in a formal sense value-less, thus gets ideological significance within its application.

Thus, the formal rationality of markets or technology does not exclude value-ladenness. On the contrary, within a specific context, formally rationally designed systems acquire specific socio-political biases.

This does not need to be only a negative feature. In Feenberg's later books (1995; 1999), he illustrates this point with far more positive examples, in some cases again taken from factory work. For example, at the end of the 19th century, factory machines were specifically built to be operated by children. As the general stance toward child labor changed, the factory machines did too. Now, we cannot even imagine that in the design criteria and process we would take into account the possibility of children operating these machines. Our cultural horizon has shifted, and our criteria for what constitutes a proper factory machine have too.

Another example is the development of steamboat boilers in the 19th century. Here, safety became an integral part of the design of these boilers after several accidents happened. Afterward no one saw these safety requirements as essentially inefficient add-ons to the boilers, while they were seen as such before.

These examples show a general feature of technological artifacts and technological systems, which Feenberg calls the "technical code". According to Feenberg, "technical codes define the object in strictly technical terms in accordance with the social meaning it has acquired" (1999, p. 87). Technical codes thus denote *specific* aspects of a collective (social, cultural) definition of an artifact or technological system, namely those that are "direct reflections of significant social values" (1999, p. 87). The "self-evident" criterion of the height of work benches actually is the social value internalized into a technical definition, that children should not work in factories. After severe political struggle in the midst of the 19th century and subsequent governmental regulation, ultimately the idea that a "proper" steamboat boiler as a "safe" boiler became internalized in technical codes issued by the American Society of Mechanical Engineers (1999, p. 95). Safety became an inherent technical criterion, instead of an imposed, but technically speaking "troublesome", add-on to an otherwise efficient design.

The reader may notice the affinity of this idea with constructivist ideas about the social and cognitive influence on design processes. Indeed, Feenberg explicitly relates this idea of a technical code with structuring concepts in constructivist thought, like technological regimes (Rip & Kemp, 1998), technological frames (Bijker, 1995), or even technological paradigms (Nelson & Winter, 1982). Although they are definitely not the same, they share the (basically Kuhnian) idea that technological change is not purely contingent, but is structured by preexisting and relatively stable cognitive states and previously attained technical standards and criteria. These cognitive and technical preconditions serve both in terms of constraints as well as facilitating factors for technological innovation. Technical codes represent specific parts of these rather broad analytical categories, namely those aspects that "can best be interpreted as direct reflections of significant social values." (1999, p. 87). We can think, for instance, of culturally dominant definitions of what an artifact is meant for, its accepted main function and cultural meaning.

This concept of a technical code thus exists in close proximity with constructivists notions of structuring frameworks. But Feenberg also needs the concept for purposes different from – but not necessarily incompatible with – constructivist analyses. As said, the prime focus of constructivism is on technical change. Dominant social values and norms do play a significant role in the way these processes of change are reconstructed and analyzed. But once these processes end, once closure has been reached, these social values are – roughly speaking – of no interest anymore for a constructivist. Change explained, job done, good night.

But just at this point, things become really interesting for a philosopher such as Feenberg. Settlement of a controversy of course does not imply that everybody agrees. On the contrary, via the creation of technological artifacts or systems, the entrenchment of the political and ideological visions that dominated their construction is increased, since they acquire a material counterpart in these artifacts or systems, counterparts that have a strong legitimating force, because the reflections of the political and ideological visions are cast in the "indisputable" logic of a seemingly neutral technical rationality the technical code. Technical codes thus settle political and ideological disputes.

A good example Feenberg gives of this settling process can be found in his *Alternative Modernity* (1995). There he analyses in considerable detail the way the French videotex system, Minitel, was conceived, constructed, and subsequently radically changed by user interventions. I cannot do justice to his historical reconstruction within the limits of this study, so I will restrict myself to the main points. Feenberg shows how this videotex system originally was envisioned by an exclusive number of actors, all linked to the tight governmental-industrial network typical for France. From their perspective, videotex was a good means for providing the public with access to official information, like government regulations, telephone records and the like. Likewise, the system was technically designed to this idea: the spreading of information from a limited number of central databases to an unlimited number of citizens. Since the French telecom company thought it was cheaper in the long run to provide their customers with Minitel terminals than to spread millions of paper telephone books each year, anybody with a telephone line could get a terminal for free. These terminals were designed specifically for database-access, for instance because the set of function keys on the keyboard had particular assignments related to data retrieval. Thus the system's constitution (the places where data was stored, the X.25 protocol for host-terminal communication, the way the terminals were designed) reflected a particular idea of what communication technology was about: dissemination of information, relevant in the eyes of a dominant hegemony of governmental bodies and national corporations. This is what Feenberg would call the technical code of the Minitel at the time of its conception and early existence. It facilitated one type of communication (information), while inhibiting or at least neglecting all kinds of other purposes and goals people might have with such a system.

Feenberg prefers cases with a happy end. Although the early days of the Minitel primarily show how dominant elites define the way electronic communication should take place, and thus to strengthen their own central roles in society, in the case of Minitel this was not the end of the story. A few hackers succeeded in by-passing the built-in limitations of the system, and constructed a primitive bulletin board system by using a host computer as a relay, rather than a database. By this reconstruction, distant terminals could communicate with each other, instead of separately communicating with a central host alone. Although this hack was observed with suspicion by the owners of the host, they did not take measures against it, since their revenues were based on the number of telephone calls made to their host computer. This was the start of a huge proliferation of similar bulletin board systems, the vast majority of which were not surprisingly dedicated to erotics and sex. The Paris walls were soon covered by posters urging bypassers to 'TAPEZ No . . .'. A new technical code was effectively created, that of communicating – instead of informating. Users thus changed the technical code of the system by subversive action.

5. A "POLITICALLY AWARE" VERSION OF CONSTRUTIVISM

The way Feenberg integrates constructivism into his own program is I think twofold. On the one hand, he adopts the constructivist theoretical framework, portraying technology as essentially interwoven with the social sphere. Likewise, he adopts the basic tenet that technological change is determined by construction processes that do not presuppose in advance one universal guiding principle, such as power or technical rationality. Therefore, technical artifacts and systems are fundamentally contingent products of historical processes. Feenberg also stresses the role social interests play in co-determining the course of technical change, but transcends constructivism by investigating the political and ideological role artifacts and systems themselves play *after* they are created.

On the other hand, he uses constructivist methodology as a way of investigating these political and ideological roles. Since technical codes reflect the social values that were dominant during the design process, one way to get a grasp of these codes is to investigate the negotations taking place during construction. Opening the black box of construction processes may have heuristic value for opening the black box of the construction itself. But, as indicated earlier, we should distinguish between these two kinds of black box analysis, since investigating the genesis of an artifact may reveal useful things about its actual constitution but it is not sufficient for knowing what the artifact is, what role it actually plays once closure has been reached.

Constructivist theory and constructivist methodology thus can be usefully applied in a philosophy of technology that strives for a critical position in addition to normatively disengaged research. Moreover, if we adopt the basic conviction that the politics of technology can be fruitfully studied only if we look for the ways technical artifacts are integrated into culture and social practice, empirical investigation of these sociotechnical complexes becomes obligatory. In my view, a philosophy of technology that exclusively orients itself to case-transcending or even universal characteristics of technology will fail to identify those features of technology that actually make a political or social difference, positive or negative. This empirical input to a philosophy of technology, however, cannot and should not be attained with an exclusive constructivist methodology, because of its fundamental focus on change rather than the construction itself. Taking up a methodological relativist stance in order to get an adequate, non-Whigghish description and explanation of processes of change is perfectly legitimate, but it should not be taken as a general norm for studying technology as a whole.

Identifying technical codes in existing technological systems and artifacts thus cannot be done with constructivist methodologies alone. On the one hand, this is due to the fact that contexts may change over time, while the artifacts or technical part of the systems stay essentially the same. The change of the political-selective role of the Long Island Expressway bridges is a telling example. While the bridges themselves have stayed exactly the same, changes in their context – the type of cars people are using, changes in economic relations – also change the selective power the bridges exert. On the other hand, the political and social constraints of technological systems may very well exceed the interests and intentions of the actors involved in its construction. (The term "side effects" seems quite adequate here, although I would use the term with great caution because of its old-style TA connotations.) Investigating their historical origin will not reveal these kinds of effects.

A constructivist analysis of the design of bridges may thus very well reveal the political agenda of one of the main actors (e.g., Robert Moses), and thus suggest (but *not* determine) a specific direction to investigate the actual political role of these bridges once they are built. This is what I would call the heuristic value of a constructivist methodology for a critical theory of technology. But once closure has been reached on the meaning and the design of these bridges, there is no special role for constructivism in investigating their actual effects, "unintended" or caused by a change in the context of these artifacts. To get insight into the technical codes of existing technological systems, we must fall back on other types of analysis. Feenberg has not as yet provided a full-fledged methodology for this, but he suggests the use of hermeneutical methods (Don Ihde, Gilbert Simondon) to highlight the social constraints of technological systems, which have been concretized into technical codes. This analysis essentially combines identification of the interests of particular social groups and the translation of these interest into seemingly neutral, or rational technical specifications and criteria. In order to formulate a technology critique of specific artifacts and systems, a good start might be to focus on those features that seem to be purely technical – or, to use an email shorthand of technophiles, to RTFM, that is "Read The F***ing (Technical) Manual".

REFERENCES

Barnes, B., Bloor, D., & Henry J. (1996). *Scientific Knowledge: A Sociological Analysis.* London: Athlone.

Bijker, W. E. (1993). Do Not Despair: There Is Life After Constructivism. *Science, Technology & Human Values, 18*(1), 113–138.

Bijker, W. E. (1995). Democratisering van de technologische cultuur [Democratizing technological culture]. Inaugural Lecture, Maastricht.

Bijker, W. E. (1995). *Of Bicycles, Bakelites, and Bulbs – Toward a Theory of Sociotechnical Change*. Cambridge, MA: MIT Press.

Braverman, H. (1974). *Labor and Monopoly Capital.* New York: Monthly Review.

Collingridge, D. (1980). *The Social Control of Technology*. New York: St. Martins

Feenberg, A. (1991). *Critical Theory of Technology.* Oxford: Oxford University Press.

Feenberg, A. (1995). *Alternative Modernity: The Technical Turn in Philosophy and Social Theory.* Berkeley and Los Angeles, CA: University of California Press.

Feenberg, A. (1999). *Questioning Technology*. London: Routledge.

Latour, B. (1993). *We Have Never Been Modern.* New York: Harvester Wheatsheaf.

Latour, B. (1987). *Science in Action: How to Follow Scientists and Engineers through Society.* Cambridge, MA: Harvard University Press.

Pinch, T., & Bijker, W. E. (1987). The Social Construction of Facts and Artifacts: Or How the Sociology of Science and Technology might Benefit each Other. In: W. E. Bijker, T. P. Hughes, & T. J. Pinch (Eds),. *The Social Construction of Technological Systems – New Directions in the Sociology and History of Technology* (pp. 17–50). Cambridge, MA: MIT Press.

Winner, L. (1977). *Autonomous Technology: Technics-Out-of-Control as a Theme in Political Thought.* Cambridge, MA: MIT Press.

Winner, L. (1980). Do Artifacts Have Politics? *Daedalus, 109*, 121–136.

Winner, L. (1986). *The Whale and the Reactor*. Chicago: University of Chicago Press

Winner, L. (1993). Upon Opening the Black Box and Finding it Empty: Social Constructivism and the Philosophy of Technology. *Science as Culture, 3*, 427–452.

PART II

ONTOLOGY AND THE EMPIRICAL TURN

OBJECT AND SOCIAL ARTIFACT IN ENGINEERING DESIGN

Louis L. Bucciarelli

In *Designing Engineers* (1994), I report on three participant/observation studies, carried out at three different firms, all engaged in the design of relatively "high-tech" products. My purpose was to develop a better understanding of the design process – to explore and determine the role of beliefs and norms as well as instrumental reasoning and experience in fixing the form of technologies prevalent today.

In the book, I present the engineering design process as a social process, a process requiring the contributions of, and collaboration among, a variety of individuals; many, but not all, engineers; each with different responsibilities and interests; each seeing the design task differently; each working, a good bit of the time, within their own particular *object world* – a world of specialized scientific theory, analytical models, instrumentation and measurement technique, dedicated bits and pieces of hardware, unique fabrication methods and (just as essential) historical precedents. But object world activity does not suffice. For designing to progress, participants must come together to negotiate, clarify, and harmonize their different contributions to the task. In these formal and informal encounters and exchanges, there is no over-arching object-world, instrumental technique available to dictate synthesis, to automatically reconcile competing (technical) interests. The process is social. The final product, the artifact, is a social construct.

In this chapter I attempt to develop this perspective in ways not considered in *Designing Engineers* and explore some further consequences. I consider three kinds of artifacts: a new, sophisticated material available to designers of

The Empirical Turn in the Philosophy of Technology, Volume 20, pages 67–80.

ISBN: 0-7623-0755-2

automobiles, aircraft, and the like; a photovoltaic module, one of the products analyzed in *Designing Engineers*; and finally a system meant to govern everyday financial transactions, albeit of a primitive kind. I am interested in testing the notion that how an object is seen – is "read", is described, is understood, has such and such (essential or not so essential) features – depends upon the interests, needs, experiences, and talents of participants in its design and use. I am interested too in how, despite the variety and incommensurability in participants' understandings, design often succeeds. In each case I inquire into the nature of the object of design, whether a basic material, a functioning product or system, hoping to provide a basis for connecting up philosophical analysis of technology in the making and what can be learned from an empirical study, ethnographic in nature.

1. A MATERIAL'S PROPERTIES

I first heard of the material *AL-MMC* at a recent workshop on the introduction of new materials into automobile design whereupon a speaker from Duralcan USA, a division of Alcan Aluminum Corporation, described its virtues. As we were "materials people" – a mixed group from the academy and industry, responsible for the drafting of a report on the state of the art of the development of innovative materials for industry – he felt no need to spell out *Aluminum Metal Matrix Composite*. After several introductory corporate slides, he did define its composition:

> *A family of aluminum alloys reinforced with Silicon Carbide (SiC) or Aluminum Oxide* (Al_2O_3) *ceramic particles.*

Now if I were in his shoes and charged with explaining this new material's features to a more general audience and wanted to flesh out this cryptic definition of the thing's composition, I might hold up a sample – an actual artifact in the form of a mechanical contrivance made of the material or simply an undistinguished chunk of the stuff – then go on to explain what is meant by the label 'metal matrix composite'. I might point to analogous composite materials – a fiberglass reinforced plastic boat hull, or a carbon filament wound frame of a tennis racquet – to convince my audience of the existence and utility of these thoroughly modern innovations. My description could be considered technical, at least at some general level.

If I passed the sample around my listeners might take some sense data themselves and estimate its density, one of the material's primary attributes. Together, we could get a feel for AL-MMC and its properties, argue about how it differs from other materials, what it might be used for and, in this way, construct (an understanding of) AL-MMC. Our shared knowledge, developed on-the-spot, based on a mix of sensory skill (the density), on metaphor (other composites), on conjecture (possible use), on history (my report about Duralcan) would enable us to carry our technical discourse further. There are other physical and chemical properties of AL-MMC I might describe and other contexts for application and use. One in particular concerned the representative of Alcan. Back to the workshop.

Moving to his next slide, he claimed "... you must put the material properties into their language." He was referring to his interactions with engineers responsible for the design of the brake system within an automotive firm – one of his potential customers and users. He wanted to show the advantages of his product and how they derive from the material's properties. His next slide:

ADVANTAGES OF AL-MMCs FOR BRAKE APPLICATIONS

- *Superior Thermal Properties*
- *Low Noise Susceptibility*
- *High Wear Resistance*
- *Lower Mass*
- *Excellent Corrosion Resistance*
- *Recyclable*
- *Viable Cost/Value Ratio*
- *Longer Life*

Leaving aside the superlative qualifiers in the above list, we can ask for an elaboration of these features and, more specifically, how are their values to be measured. The response for some would appear to be straightforward and stand independent of any particular social context. For example, a posited value for the material's mass density would gain consent with little fanfare, assuming the competence of the person doing the measurement using balance beam and meter stick. The material's thermal properties – its thermal conductivity, its specific heat, its coefficient of thermal expansion – would require a bit more attention to standard methods of measurement laid out in some reference book authorized by an engineering society but these too might be considered primary properties.

Others on the list, however – e.g., noise susceptibility, wear and corrosion resistance – are not so simply defined, nor measured. For these we must go to the American Society for Testing Materials Standards (1996) to search for appropriate means to establish universally applicable, quantitative measures.

For example, in volume 03.02 we find several tests prescribed to measure wear resistance of metallic materials: There is a

Test Method for Galling Resistance of Materials (G 98);
Test Method for Linearly-Reciprocating Ball-On-Flat Sliding Wear (G 133);
Test Method for Ranking Resistance of Materials to Sliding Wear Using Block-On-Ring Wear Test (G 77).

Note: The number in parentheses refers to the identity of the group responsible for drafting the prescribed method.

For corrosion resistance we find, among a dozen or so possibilities:

Test Method for Determining Susceptibility to Stress-Corrosion Cracking of High Strength Aluminum Alloy Products (G 47);
Guide for Conducting Exfoliation Corrosion Tests in Aluminum Alloys (G 112).

A closer reading of the text reveals the conventional nature of a property measured according to one of these prescribed testing procedures. Their multiplicity alone reflects the importance of context in establishing quantitative measures. We might say that the value of the property depends upon the observer's point of view or upon a designer's interests in the material's potential application. As such, we might label these attributes "secondary". Yet reflecting back upon our more primary properties, e.g., mass density and specific heat, we acknowledge that even these are conventional in some sense, dependent upon a stabilized intellectual as well as instrumental infrastructure for their final and formal definition.

This foray into the world of standard testing procedures is meant to provoke thinking about just what a property of an object is, how it is defined, its ontological status. If we claim the material "contains" its properties in itself, independent of any potential social context of use, we must acknowledge that the definition and measurement of any particular property does require consensus and agreement about the context, means, and methods of measurement: Certain instruments must be deployed, the material specimen must be given a certain shape, environmental variables like temperature, humidity, atmospheric pressure, local acceleration due to gravity, electromagnetic field strength, vibration levels, dust content, must be controlled, artisanal skills must be brought to bear to obtain a "good" reading. All of this activity is done in accord with the conceptual apparatus and dictates of some scientific theory. The quantitative value of a property is only fixed through interaction with and within this sort of tightly controlled, theoretical, methodological, and physical setting.

Is a property, then, a social construction? If I limit my meaning of social to apply to a process of immediate human exchange within a well-defined group, e.g., the participants in a design project within a particular firm, working on a particular design task, then to answer yes is a bit overblown. But, if we allow a more open reading of "social", properties are evidently social constructs, albeit in different ways. Some, like *Recyclability or Cost/Value Ratio* may be locally acknowledged, defined, and measured within a firm or industrial group. Others like *Corrosion Resistance* or *Wear Resistance* are fixed, as we have noted, within formal social processes by professional committees; properties of this sort are truly "conventional", the result of a convening of interested parties and proclamation of measurement technique. Others, like mass density, specific heat, or dielectric strength appear to stand independent of any such deliberations. But they too are social constructs in that, as the history of science reveals, they are the product of informal exchange and the weaving together through historical time of multi-colored threads of understandings about scientific principles and concepts, about experimental methods and results, into a binding knot. In all cases we see how dependent upon contingent, historic processes the attribution of a material property can be.

Apart from any philosophical significance the evidence I put forward may have, it is important to address these questions if one seeks a better understanding of how objects (materials) are adopted, manipulated, reconstituted and put to use in engineering design. For in product design and development, where the context is not congruent with that of property measurement, participants in design must make a mapping of the material's attributes onto the design and attempt to project their influence on performance. Some properties enter into their analyses explicitly as parameters, others not so directly. In either case their effect and/or importance is often not explicitly or immediately apparent. Modeling and simulation provide readings but these models are in themselves complex and reveal the effects of property variation only through multiple runs. While the aim of the ASTM standard testing procedure is to isolate the property, make it intrinsic, like mass density, and so available universally for exploiting in the design of who knows what range and variety of products, the context of design within any particular firm differs from the context of property measurement.

The point the representative from Alcan is making when he announces "you must put the material properties into their language" is that it is no small challenge to define the attributes of his new material in a way automotive designers could understand as relevant to their domain of expertise and interest. This singular technical and social context requires him to build upon his listener's experience with existing brake system materials. Just as I, at the outset of this section, described how I might build upon an audience's experience with materials, he must work to establish an identity of the material in the automotive design context; to move his potential customers to incorporate the material into their design thinking. Ideally, these properties will become part of brake designers' vocabulary, assuming they are convinced to use the new material in their next designs. Just as mass density has its place within a Newtonian analysis of the dynamic forces and stresses engendered in a rapidly rotating brake drum, so too measures of corrosion resistance might find their place in the engineer's discourse about brake system reliability and lifetime.

So what is AL-MMC? My claim is that only within particular social contexts are the material properties, and hence the material, usefully defined. Any response to the question requires an elaboration of context including a definition of properties of interest and how they are to be evaluated.

2. A PRODUCT'S COMPLEXITY

The making of a "new" material, a composition fabricated of more rustic stuff, requires thought, planning, testing, trial and error – in short, AL-MMC is itself the result of a design process. The object I pass around might appear as mundane and commonplace as any ordinary stone lying about, but looks are deceiving. Because it, in itself, doesn't do anything but must be further shaped, thought about, manipulated, machined, tested, and then reworked before we encounter it in a product, our valuation of the chunk I circulate as a technological innovation is liable to be limited. We note its potential as a resource, a better building block, for making things like automotive braking systems that apparently do *do* something, but discount its own complexity.

AL-MMC is, in fact, multi-faceted, multi-dimensional. As an outsider to the material making/using process, we see only two or three of its faces; we have some inkling of its mechanical properties (we have measured its mass density), but ordinarily remain uninformed of its thermal characteristics or other dimensions fixing electromagnetic properties for example; or still others bearing on financial, aesthetic, maintenance, or ecological concerns – dimensions spanning the design space of automotive brake systems. We don't see its high wear resistance – indeed, we do not even directly sense its mass. Certainly we don't see the process of its design and development nor can we fathom all the ways and schemes through which AL-MMC might become a constituent of recognizable technological productions. In particular, which properties might be considered most valuable (primary?), which not so (secondary?) in any future product development – if we were made, or could be made, aware of them all – is hidden from view.

When we turn to the design of things that *do* do something, another kind of deception prevails. Since AL-MMC is amorphous in appearance, we are hard put to respond if asked – what does it do? We might rattle off a few of its properties, but they don't do anything in themselves alone. The intended purpose of a brake system is, in contrast, quite clear: It stops the car when engaged. Or, drawing upon one of the three studies included in my book, a photovoltaic module produces electrical power directly from the sun. If this is so, if the answer flows so easily, where is there deception?

If I were to pass around a photovoltaic module, certain attributes would immediately come to the fore. It would appear as a panel of rectangular shape, size, and weight of human scale (which makes its passing around a real possibility). The front which faces the sun appears as an array of usually circular, deep blue disks, protected from the weather by a plate of glass and rimmed with a slight frame. The disks must be the important ingredients in the production of electrical power because they are all connected by thin, flat wire in a network. It makes an attractive picture. Turn it over and the most distinctive feature is a small box which, upon opening, reveals a red terminal screw marked "+" and another black terminal marked "–". That is evidently where you hook up the module to draw off the power.

The object is more than a size, a shape, a color patch, with geometric appendages; it's also a voltage, a current, a power in kilowatts; an electrical network of cells in series strings and parallel; it is a receptor of solar flux to be oriented at the appropriate tilt; it's a sandwich construction that must seal out the weather; a collection of materials that must resist corrosion. It's measured in all of these dimensions.

Up closer still, the module's properties have a further dimensionality; "seeing" requires more than the impression of an image in the mind's eye or familiarity with some performance oriented, engineering features. Can you see the electron holes migrating through each individual cell? Can you see the flexing of the frame in the wind and the bending stresses engendered in the glass? Can you see the electrochemistry of the corrosion of the leads and terminals within the junction box as the years pass, the reduction in power due to mismatch in the individual cells' short circuit currents, how the maximum power point shifts as the temperature of the module changes, the kilowatts of radiant power per square meter that impinge upon its face and how that changes with latitude, local atmospheric conditions, and the seasons?

Like AL-MMC, a photovoltaic module is multi-faceted; as an outsider, we see but a few of its faces. And so too, when we turn to ask how it came to be, the deception takes on a different guise. It's not just a matter of not seeing, but rather we, in a sense, see too clearly. Standing apart from the process of design, and in retrospect after the process is over, we are tempted to analyze the logic of operation of the photovoltaic module and read out how it came to be; we impose the logic of operation upon its design (Bucciarelli, 1994, p. 14).

> The artifact is a rationalization of itself, one that excludes alternative forms and speaks to us thus: "I am a working, efficient, marketable machine. Knowing how I work, understanding my underlying form as the scientific principles that govern my doings, and reading my documentation . . . you can reconstruct the decision-making process that made me (or rather that allowed me to make myself.)

In this way we claim that the product, seen as a harmonious whole, is but one systematic, instrumental, coherent solution to a singular challenge. It is so hard to see (back) otherwise; to see the different views, evaluations, conflicting arguments, lengthy negotiations, and incompatible positioning that designing happens to be. This is the deception of ready appearances, akin to the pull of the present on the writing of history, which is particularly acute in doing the history of science.

3. AN OBJECT'S WORLD

As the object is more complex and compounded than apparent at first glance, so too the design process is complex and multi-faceted, mirroring in its organization the product's multidimensionality. How the object of design is perceived and described by participants in the design process depends upon whom you encounter. An engineer from the panel fabrication group sees and describes the different layers that make up the module in cross-section; to him the way these layers are joined, how the whole is sealed and fitted within the frame and the strength of the frame constitute the object. To the materials specialist responsible for quality control of the product's electrical performance, how the cells are selected and networks together constitute the module; to her the electrical efficiency and how this depends upon the integrity of the cell growing process is uppermost in her mind. The engineers in marketing and systems engineering, on the other hand, see the module as a "black box", as a closed unit which, when combined with others exactly like it and set out in a field or on a customer's rooftop, produces a certain amount of power at a certain cost; to them the behavior of the entire system – solar resource, battery storage, controls, and load as well as module – constitute the object. What are essential properties to one are accidental to another.

Each participant in designing sees a different object according to his or her technical interests and responsibilities. I say that each works within his or her own respective *object world*.

Within their respective domains there are, not just some particular laws of nature, concepts, and theory, but well known instruments and standard testing procedures. Just as important are accepted modes of description and argument, a specialized vocabulary and variant ways of representing the object and its behavior in a diagram, in a sketch, in a computer model. There are standards too, and formal and informal norms and rules. The formal, like the ASTM prescribed test methods, are but a small piece of the action; other rules, usually un-articulated, guide designers working within object worlds just as surely. What is a good argument, deduction, inference of possible cause? When is one cutting corners, when over-designing? What is a reliable test procedure, when have you taken sufficient data? When do you have a "fix" on a bug?

Object worlds are places for hard instrumental thinking and doing grounded in science and professional traditions. Participants' foci may vary and their thoughts be bounded differently in different object worlds, but each can make claim to empirical certitude. And this is what makes the design process social since there is no over-arching, object world whose instrumental methods might serve to reconcile the conflicting and competing proposals of different participants.

Are these object worlds "incommensurable"? One reviewer (Constant, 1996) described them as "incompatible". Certainly I mean to suggest something of the sort, at least in the sense of their paradigmatic status as empirical practices. I claim that there was no over-arching instrumental technique, no super object world, that might reconcile all the differences in vision, proposals, and prescriptions. This is the basis of my claim that designing is a social process, a process of negotiation.

Kuhn (1962) makes a strong and difficult case for incommensurability of Newtonian mechanics and Einstein's revision. Standing apart from the thick of the battle, we are tempted to see the former as a special case of the latter (for velocities small with respect to the speed of light). He posits, and its so hard to accept, that a Newtonian of the 18th century saw the world differently from an Einstein of the 20th century, that they lived, so to speak, in different worlds.

But there is a difference with Kuhn: For while object worlds *are* incommensurable in that they are rooted in different scientific domains and they are in competition in design – the structural engineer demands adequate strength, the propulsion engineer demands minimum weight; these pull in opposite directions – unlike Kuhn, the competing claims of participants from different object worlds can all be true and coexist within their respective domains. It is not a question of replacing the claims, the way of seeing, or the theoretical basis of one object world by those of another, as the Copernican world replaced a Ptolemaic world, but rather the task for the inhabitants of object worlds is to creatively find a way to get around and/or build upon the tension.

What is essential to grasp is, again, that there is no over-arching instrumental method for doing this, one that can reconcile the conflicting and competing claims generated within different object worlds. (Trade-off is the cover term that hides the complex, multifaceted, and far from strictly instrumental process that leads to the construction of binary oppositions, seemingly able to be evaluated in instrumental terms.) This does not mean that "anything goes", i.e., that participants in design, when off and away from their home ground, can fantasize and construct as if there were no rules at all. No, in designing there are always constraints – scientific laws, organizational forms, infrastructural resources, corporate norms – which guide as well as restrain participants. In this, it is again akin to Kuhn's vision in that social and cultural context does matter.

4. OBJECT AS SOCIAL ARTIFACT

This perspective on a material and properties, on object and object worlds, on design process as a social process, has certain consequences. In a sense it robs the object as artifact of its traditional identity as machine or system working according to scientific principle devoid of ambiguity, uncertainty, or even human agency in its making; or, if you prefer, its identity remains underdetermined unless one pays specific attention to the social contexts of design and use. Does it make sense, in any but the most general and abstract contexts, to speak of the object in itself alone, unencumbered by any human act, thought, or discourse?

The identity of an artifact, the way it can take on different meaning within different contexts is already signaled by those who speak of the artifact as a "black box" (Latour, 1987). This is a useful concept that recognizes how any technological production can be viewed and put to useful purpose without knowledge of its inner workings or underlying form.

For example, at the firm designing photovoltaic modules, the systems engineer could, at times, treat the product in this way. The science of its individual cells, the details of the electrical circuitry, mattered little as long as the module performed as designed, that is, produced the correct current as a function of voltage at the anticipated solar flux. As long as the current and voltage "seen" at the output terminals were as predicted, given the level of intensity of the sun's radiation impinging as input, the module could remain closed. Yet at other times the same engineer might very well be concerned with the finer details of the module's performance; e.g., at a meeting to fix the nominal module voltage, in consort with design participants from other object worlds, the black box is opened up and the circuitry and individual cell characteristics become of immediate concern. At another meeting within the firm, the individual cell might be "opened up" and its underlying form closely inspected and negotiated. What then is the black box depends upon context and the interests of those who are concerned. We might say that the object appears as a black box of potentially infinite layers and levels.

Black box, as the concept is most usually employed, refers to some stabilized, though not necessarily physical, apparatus. In design, in contrast, when the object does not yet exist as hardened artifact, we can still speak of the object as having an identity but we are forced to construct that representation out of design participants thinking and doing. Identity then is of a collective sort. Like a social fact in the sense of Durkheim, the object exists as a *social artifact.*

I digress to describe a social phenomenon that has the characteristics of a social artifact as I mean the phrase – a product whose design and development is not under the control of any one individual, that no single person understands completely, yet all contribute to in the making.

A phenomenon one observes in the United States is what I call the *emergence of penny pools.* A penny pool is the little reservoir of pennies, found aside the cash register in many, but not all stores, usually in those run by smaller merchants. The pennies, numbering anywhere from a few to several dozen, lie in a small open container, roughly the size of an ash tray (which may indeed be an ash tray) on the counter, aside the register. Some are purposely made for commerce and will show an advertisement, e.g., for cigarettes, with the text "take a penny, leave a penny". But just as many are simply makeshift containers of the store management or cash register attendants own making. Why are they there?

A utilitarian, rational choice theorist would explain their existence as follows: People do not wish to carry pennies around; furthermore, the cost of transaction – taking or giving the pennies in change, putting them in the purse or pocket, or searching for the small coins in making payment – is not worth the one to four cents the coins represent. Hence customers and store owners, as rational actors, have agreed to setting out a penny pool as a reservoir to be tapped when the customer needs to pay off one or two cents or when the store owner needs to pay out one or two cents. When the market basket price modulus 5 is 1 or 2 cents, the customer "rounds down" to the nearest five or ten cents in making payment from his or her pocket, then takes 1 or 2 cents from the penny pool to make full payment. When the market basket price modulus 5 is 3 or 4 cents, the customer "rounds up" to the nearest five or ten cents in making payment from his or her pocket, and the store owner puts the 2 or 1 cent due back in change into the penny pool. In this way, the penny pool does away with the need for customers to carry pennies about on their persons.

Here then is a social artifact, a system for the conduct of commerce that appears to have no single author, no government agency dictating regulations, a truly free market phenomenon whose design is the result of a collective recognition and shared system of values – about the worth, rather worthlessness, of the penny.

Yet how did this happen? How rational a design process might this be? From atop the mountain where our rational choice theorist sits, the questions seems irrelevant – the theorist has an explanation, a model, a theory which is played out in the terms of the relative value of the penny to the customer and the mathematics of Markov Processes which predict the stationary states of the "depth" of the penny pool, given some assumed market basket price distribution and the probable number of goods in the basket. The theorist looks at the final design and constructs an instrumental explanation that not only describes how the system works, but also implies the rationality of its design.

If, however, you go into the field and ask customers, store managers, and cash register attendants why this pool of pennies is there you will get a range of responses, some contradictory, some nonsensical, some of which fit with elements of the reasoning imputed to individuals in a utilitarian model, but none fully consonant with the way the pool should be used in a fully rational way, namely as a means of rounding up or rounding down that eliminates the need for the customer to carry pennies. For example, some pools are only used when pennies are given in change to the customer, rarely tapped for pennies by the customer; the proceeds are thought to go to charity; customers take but one penny at most from the pool; the pool is kept hidden by the store owner, only tapped upon demand by the customer; many customers, or even register attendants, don't know why they are there; indeed, even the formally "designed" receptacles reveal a deficiency; they should say "leave one or *two* pennies, take one or *two* pennies". All of this does not disprove the rational choice theorist's model but it does place limitations on its implications. The rational choice theorists can treat individuals as atoms, devoid of reasoning, and well they might since most individuals will be judged to be behaving irrationally if asked! But it should be made clear that this is the perspective adopted; the model is not about "real" individuals but about an aggregate calculating individual customer in exchange with some aggregate accommodating individual attendant.

Penny pools are a social artifact, reflecting shared values at some level – all would agree that pennies are worthless and not worth haggling over – but how this general concern moved participants in their production and use is not so simply explained. Evidently the process that brings them to us is more complex than that implied by any utilitarian theory. (The passive voice is appropriate in that it is difficult to define a single author, agent, government bureau, or chamber of commerce promoting the system.)

Of course the engineering design process is not this haphazard but it does exhibit some of the same qualities. When you look from afar at a finished product – a chunk of AL-MMC, a photovoltaic module – the "inherent" properties or the instrumental machinery of the product or system guide our thinking about and explanations of how the thing work. When we address and describe the design process itself, these again dominate; the identity of individuals and their different contributions, responsibilities, and often conflicting perspectives on the product remain hidden, seem to be of secondary importance. Just as personhood enters solely as some faceless bearer of rationality in a model of penny pool transactions, so too the designer in many models of design process enters only as some nameless (and genderless) agent responsible for the application of well established scientific principle and instrumental technique.

It is only when one attends to the finer details of process, and makes an empirical turn to engage the day-to-day activities of participants, that another image emerges. Only then does one see the negotiation among individuals with different perspectives on the task, the trials, the errors, the creative resolution of competing, if not conflicting proposals.

5. CONCLUDING CONJECTURE

I have argued for a relativistic reading of technology, one that attempts to free the mind from the rigidity and lawfulness of the technical production as artifact standing alone – whether that be a new material, a photovoltaic module or, even a financial accounting system. I have encouraged the reader to come down from the mountain and "see" the social entanglements in design and in use as a necessary dimension to that reading.

I have spoken of object worlds and the individual's activity in a one-on-one engaging of the object of design. The use of the qualifier "object" ought not to bias our reading of the features of these worlds. Of course participants in design, working within object worlds, work with things; some are more cluttered than others with bits and pieces of hardware, a rapid prototype, a piece of last year's model, a component being readied for proof testing. But the object of design in a holistic sense does not exist as thing. What does exist and is primary, over and above things, are theories and images, histories and exemplars, heuristics and special modeling methods, ways of judging and ways of speaking, tacit as well as documented knowledge – peculiar to the individual's discipline that help define the product's performance relative to the particular concerns of his or her world. Perhaps I should say "subject of design" rather than object of design.

But designing requires more than object world work; the design review meeting, the conference call with a subcontractor, the informal three way decision-making chance meeting in the hallway, and numerous other occasions for social exchange are just as essential parts of the process. In these occasions for transactions across object worlds the object of design becomes something more than the focus of specialized apparatus, theories, techniques within a particular object world, more than the intellectual property of any single participant. Here the object exists in a collective sense, as social artifact. The community has it in mind all the while; the collective maintains it, keeps it energized in their collective memory.

The implications of this perspective, if it has any validity, are significant. In my book, I describe some of the difficulties of evaluating the quality of a design if we accept this vision. A related problem arises if the boundaries separating the narrow technical ingredients of a design from its social and cultural milieu become diffuse, as they will if we accept the claim that there is no over-arching, object-world, instrumental method or theory able to reconcile conflicting technical demands. Where does the technology stop and society begin? What happens to "impacts" if these boundaries disappear? More serious, how do we affix responsibility for design poorly done? How do we judge negligence of a collective enterprise, if we acknowledge that no single participant has complete command and knowledge of the workings of the product, of the entire system?

Finally, how do we theorize about process? Is thick, ethnographic description the best we can do? All we need do? If so, how do we draw out of this story's recommendations for action, for say improving design process? Who do we tell? How do we change a social fact?

REFERENCES

American Society for Testing Materials. (1996). *American Society for Testing Materials Standards.* Philadelphia: ASTM.

Bucciarelli, L. L. (1994). *Designing Engineers.* Cambridge, MA: MIT Press.

Constant, E. Untitled Review, *Isis, 87*(1). (March, 1996), p. 205

Kuhn, T. (1962). *The Structure of Scientific Revolutions.* Chicago: University of Chicago Press.

Latour, B. (1987). *Science in Action.* Cambridge, MA: Harvard University Press.

THE RELATIONAL ONTOLOGY OF TECHNICAL ARTIFACTS

Anthonie W. M. Meijers

In the introduction to this volume, it has been argued that an empirical turn is necessary in the philosophy of technology. Philosophical reflection should be based on empirically adequate descriptions of technology. In addition, it was suggested that the philosophy of technology should not restrict itself to normative issues, but should include methodological, epistemological, and ontological questions concerning technology. The combined effect of these recommendations is that a whole range of exciting questions will move to the center of the philosophy of technology. These questions address the internal structure of technology (design, operational concepts, functional composition), but also its relation to actual or potential users and their intentionality.

My aim in this chapter is to contribute to an ontological analysis of technical artifacts, i.e., of artifacts designed and developed by engineers. I will not be concerned with the epistemology of technical artifacts. There is an obvious relation between the two, since we need epistemic criteria for our ontological claims. My analysis focuses, however, on the basic ontological properties that technical artifacts have. I will present arguments for the idea that they are relational entities. They supervene not only on their physical structure, but also on their wider engineering and social contexts. If we want to say what an artifact is we have to specify not only its material structure, but also the practice of its design and use. That may seem trivial to some, but is not when it comes to actually stating these two very different but *interdependent* aspects of artifacts. A central part of my argument focuses on functions and functional properties. These are realized by the physical structure of the artifact together with the practice of its

The Empirical Turn in the Philosophy of Technology, Volume 20, pages 81–96.

ISBN: 0-7623-0755-2

design and use. The idea of artifacts as relational entities implies that it is not immediately evident what the boundaries of artifacts are: where does the artifact begin and where does it end? This question becomes especially relevant if we want to make an empirical turn in the philosophy of technology, and base philosophical analyses on adequate descriptions of technology. What could be an adequate description if artifacts do not have clear boundaries?

In Section 1 I start my ontological analysis with the various kinds of properties that artifacts have. In Section 2 I discuss the notion of technical function. I will defend the idea that ultimately this notion has strong conceptual links to the notion of human action and relates artifacts to the actions and practices of users and designers. A function is a *relational* property and is distinct from such intrinsic physical properties as mass or length. Relational properties, so I will argue in Section 3, are ontologically not reducible to non-relational properties. The conclusion of my argument is that technical artifacts belong to the same ontological category as social artifacts, such as laws or driver licenses. There is, however, an important distinction between the two, related to the role of physical structure. For technical artifacts the physical realization partly *explains* the function of the artifact, whereas for social artifacts it does not. Their physical realization is to a large extent arbitrary. Based on my analysis I will further explore in Section 4 the idea of an empirical turn and the proper place of case studies in the philosophy of technology. A conclusion follows in Section 5.

1. TECHNICAL ARTIFACTS AND THEIR PROPERTIES

A suitable starting point for my analysis is the observation that from an ontological point of view a technological artifact is not simply a physical object. It is, of course, a physical object, but we will miss many of its characteristics when we commit ourselves from the outset to the simplistic and reductionist view that a technological artifact is simply a physical object[1]

A technical artifact is at the very minimum a physical object that is the *result* of intentional behavior. It is made or used, *and* it is made or used for some purpose. Randall Dipert (1995 and 1993) makes an important distinction between instruments and tools, to bring out the various intentional elements involved. An instrument has the characteristic that at least one of its properties is believed by someone to be a means to an end and has been intentionally employed in this capacity; for example, a stone that is used as a paperweight. A tool, on the other hand, is an object whose properties are intentionally modified to make it a means to an end, and whose properties are believed to be so by its user. Thus, a tool not only requires beliefs about the artifact, but also beliefs about other minds, and about other agents' beliefs and intentions with respect to the artifact. An overhead projector, for example, is such a tool, requiring beliefs about the object and beliefs about the intentions of its designer or its maker. As such, tools form a subclass of instruments.

Dipert's third category concerns artifacts. They are regarded as tools that intentionally *signal* or indicate that they are tools. Artifacts signal that they have a certain purpose or effectiveness. They advertise themselves as such, so to say. This is related to the idea that for a tool to be a tool it needs to be recognized and believed to be a tool. If it is not recognized as such, its effectiveness and function is lost. User-friendly design is about this signaling. The user is guided naturally, i.e. without further documentation, to the function and use of a particular device.

The properties of technical artifacts can be classified ontologically in at least three different types. These types are: (1) intrinsic or observer-independent properties, (2) context-dependent properties, and (3) relational properties.

Technical artifacts, being physical objects, have a number of properties that are independent of observers or context. These may be said to be *intrinsic* to the system. Take, for example, a central heating unit for domestic use. Once it is built, the dimensions of the unit, the number of burners, the physical processes underlying the heating up and flow of the water in the tubes, the turbulence in the flames, the condensation of water in the exhaust gas, or the electric current in the thermostat, do not depend on the mental attitudes taken by users, buyers, sellers or, more generally, observers. They can be regarded as intrinsic, non-contextual and non-relational.

However, not all the characteristics of an artifact considered as a physical object are intrinsic. Some depend on their wider context or environment. The overall efficiency of a refrigerator, for example, depends not only upon its internal make-up, but also on the temperature outside the unit. Or, to give another example, the brake power of a car depends not only on the intrinsic features of its brake system, but also on the properties of the road surface. On a dry road a car has a better brake power than on a slippery road. This context dependency is to be distinguished from the next type of properties: relational properties.

Whether or not a property is to be regarded as intrinsic or context-dependent depends on the way the system and its boundaries are defined. Larger systems obviously include more contexts. Systems are often defined relative to our cognitive interests. For example, if our interest is in heat transfer processes, a system may be defined that contains a water pipe and a flow of exhaust gas around the pipe. If our interest is in the thermodynamics and energy efficiency of domestic heating units, this system will be a subsystem of one which also includes part of the environment of the heating unit.[2]

In the description of an instrument given above an important characteristic was mentioned. For an instrument to be an instrument at least one of its properties has to be *believed* by someone to be a means to an end, and also that it has been intentionally employed in this capacity. Tools have in addition the characteristic that they are – and are believed to be – intentionally modified to function as a tool. These properties are *relational* properties. They exist only in relation to designers, users, or observers who believe that something is a tool, intentionally modify it or use it as a tool. They are mind-dependent properties, i.e. properties that require for their existence human minds and consciousness.

The property of intentionally signaling that something is a tool, which I mentioned earlier and which is part of Dipert's notion of a technical artifact, is also a relational property. Communication or the use of signals is always directed at an audience and exist only in so far as it can be grasped by this audience. If nobody is able to pick up the message, there will be no message and therefore no communication. A signal is only a signal relative to a practice of using signals.

2. FUNCTIONS OF ARTIFACTS AS RELATIONAL PROPERTIES

So far I have only implicitly mentioned the important category of a function. Artifacts are made and are made for some purpose. That is, they have a function, which may be either the intended function by the designers or the actual function for the users. A refrigerator has the intended function to keep food or beverages at a temperature of about 4° Celsius. Its actual function may be like that, but it can also function as a bookcase or even a device for heating a room.

The notion of a technical function is not very well developed in philosophy. Most theories of function have a background in the philosophy of biology. Arno Wouters (1999, pp. 8 ff.) distinguishes four approaches to functions and functional explanations in functional biology:

(1) *Etiological theories*. According to these theories, function attributions specify the effect for which a particular trait was selected in *the past*. These effects are called by Ruth Millikan the *proper* functions of that trait. For example, hearts have the proper function of propagating blood, if they were favored by past natural selection because of that effect.
(2) *Survival value theories*. These theories identify functions with particular effects that *currently* make a contribution to the survival and reproduction of an individual organism.
(3) *Causal role theories*. In this approach functions refer to the role of certain parts or activities in maintaining a complex capacity of a system. The forked tongue of a snake, for example, contributes to the snake's capacity to find prey.

(4) *Design theories.* These theories relate the presence and specific character of certain traits or behavior to the make-up of a system and its environment. They seek to explain the usefulness of these traits and behavior in terms of the survival of that particular system in its environment.

These theories of function are not easily transferable to the technical domain. Beth Preston (1998) explicitly discusses this issue in a recent article. She questions what she perceives as a common assumption among philosophers of biology, viz. that their account will be useful for an analysis of artifacts. Preston argues for a pluralist theory of function, one which contains at least two notions of function: a notion of proper function (cf. Millikan, 1984) and a notion of system function (cf. Cummins, 1975). Proper functions are explained diachronically in terms of the selection history. Every organ or system has functions that are biologically proper to it, because they contributed to the survival of its ancestors. System functions, by contrast, are explained synchronically and they refer to the role particular components of a system play in realizing the capacities of that system as a whole. The notion of a proper function allows for the distinction between functioning, on the one hand, and malfunctioning or non-functioning, on the other, whereas the notion of a system function does not. There is no function that a component *ought* to have in the latter conception. In Preston's view, any theory of function which does not include *both* conceptions of function is incomplete and not able to account for the variety of functions biological creatures have. The notion of system function is necessary for understanding the specific characteristics of exaptations, i.e., the use of traits of individual organisms in ways they were not designed or selected for.

A pluralist theory of function is able to account for artifacts, though it will need revision and extension to cover all relevant aspects, according to Preston. Artifacts do have proper functions. These are the functions we usually refer to when we ask "What is that for?" They acquire them in a process basically similar to natural selection. They are first developed, tested, and eventually adjusted. Thereafter comes the phase of production and, if the artifact and its function are successful, of reproduction. The artifact will be selected and survive in the market. Though Preston will probably be among the first to notice that this analogy fails in certain respects, she defends it against a number of objections.

The main objection is that artifacts are the result of rational planning and intelligent thought, while "biological innovation" is the product of purely mechanical forces. Variation and selection are indeed blind processes in the biological evolution. This is a formidable objection, though Preston believes that rational design does not constitute a decisive difference. Design and rational planning increase the likelihood of successful performance and thus reproduction, but not necessarily so. Many artifacts ultimately fail. Their success not only depends on the process of their making and whether they perform as intended, but also on their acceptance in the market and the existence of competing products. So "although intention and rational planning may speed the process up and give better assurance of improved viability at any specific step, the difference [between artifacts and biological organisms] is really one of degree, not of kind" (Preston, 1999, p. 245).

Preston is certainly right that the success of artifacts does not solely depend upon good design and planning, but I believe that her answer is fundamentally unsatisfactory. Processes of variation and selection can always be regarded as similar if viewed on an abstract level in which they appear as black boxes with particular input and a certain output. These processes result in biological traits with proper functions in the biological case, and proper functions of artifacts in the artifactual case. But once we open the black box we find processes of variation and selection that are very different. It is even unclear whether the notions of variation and selection can be applied literally to the domain of artifacts.[3] Variation in evolution is a blind process in which genetic variations result in biological structures that turn out to have certain functions in a natural environment. The design of artifacts, by contrast, is an intentional process in which a desired function is the starting point and a certain physical structure together with a user manual the outcome. There is not much similarity here. The biological notions of selection or survival also do not appear to have literal equivalents in the artifactual domain. The term "selected" is already an intentional description of what is in fact a non-intentional, mechanical process in the biological domain. Artifacts are selected by human actors in the sense of being used by them for a desired function and, if successful, continued to be used for that purpose. This selection process is intentional and guided by a variety of reasons. It is not simply a causal process as in the biological case. In addition, economic survival is not to be confused with survival in terms of fitness in a particular environment. Economic survival has often very little to do with fitness to perform a certain function[4]

Another asymmetry not to be overlooked here is that the function of an artifact is explained by appealing to both its physical structure and to human actions and their contexts. A screwdriver, for example, realizes its function on the basis of its physical structure and of human actors able to use this structure in the appropriate way in the relevant context. What is proper to a proper function of an artifact is therefore established in a way different from the biological case. Biological functions can be explained solely in terms of the genetic makeup of the trait and the biological environment in which the organism is embedded. They do not refer to the activities of other agents.

In addition to proper functions, Preston argues that artifacts also have system functions. These are functions that an artifact acquires in its lifetime, though it was not designed and built for that function. We stand on chairs, use knives as screwdrivers, and so on. Some of these system functions become part of a common cultural practice, while others are highly idiosyncratic. Artifacts do not lose their proper function because of their system function. They can easily lose their system functions though. At the same time, a system function may develop into a proper function, once it becomes explicitly designed for that function. This shows, in Preston's view, that there is a dynamic between the two types of function, and that both types are necessary to explain the actual functioning of the artifact.

Whatever may be true of the analogy between the biological and artifactual domain, the discussion of Preston's pluralist theory shows in my view that we do indeed need at least two notions of function to account for technical artifacts. Here I agree with Preston. We need a "proper" function, related to the design of artifacts and to the intentionality of designers. But we also need a "system" function, related to the particular use of artifacts that does not accord with the function for which the artifact was designed.

Both notions of technical function are thoroughly relational, which brings us back to the starting point of our discussion. A proper function and a system function *relate* an object to a human *practice* of design and use. This shows a further point: the notion of a function is an action-oriented notion. The ontological category of human action and human intentionality is presupposed in any account of technical function. In case of complex systems a technical function is also relational in that it relates a functional object to the system as a whole. The functional decomposition of complex systems shows the dependency relations between the functions of the various components and the function of the system as a whole.

Ontologically speaking, artifactual functions supervene on both human intentionality and the physical make-up of artifacts. You cannot sit on a chair that has the wrong shape or is made from the wrong material, while you also cannot communicate with telephones if you don't know how to use them. There is underdetermination or multiple realizability here in two related ways. In a physical way, in that the same function (being a chair) can have several physical realizations, while the same physical structure (a wooden object) can have several artifactual functions. There is also underdetermination in an intentional way, in that the same function (long distance communication) can be realized on the basis of several human practices (mobile telephoning, waving flags), while the same human practice (waving flags) can make possible several artifactual functions (signaling, ventilating).

From my discussion it will be clear that the notion of an artifactual function is in need of further clarification. I have only sketched an elementary distinction to be made in the theory of technical functions and argued against a close analogy between the biological and technical domain. There are very many questions, however, that are still unanswered, or even not yet asked or not asked in the right way. These concern the very notion of technical function itself, the taxonomy of technical functions, the distinction between simple and complex functions, functional part-whole relationships versus structural part-whole relationships, the usefulness of the distinction between proper and accidental function, the relation between function and design, the normativity of functions and its origin, the epistemology of functional knowledge, the logic and formalizibility of functions, etc.[5]

3. THE IRREDUCIBILITY OF RELATIONAL PROPERTIES AND ITS CONSQUENCES

The previous sections result in an ontological account of artifacts as entities with at least three kinds of properties: intrinsic, context-dependent, and relational. I now want to address the more general issue of reductionism with respect to these *kinds* of properties. Such a discussion is important, because if it turns out that ultimately all kinds reduce to a single kind, little will be accomplished by the analysis from an ontological point of view. But if it can be shown that artifacts have irreducible relational properties apart from their intrinsic and context-dependent properties, the issue of what *constitutes* an artifact cannot be answered independently and separately from its wider context.

The question of the reducibility of relational properties has been a recurrent theme in the history of philosophy. With hindsight we may say that the Aristotelian ontology of substance and (non-relational) attribute has been a real hindrance for developing an adequate understanding of the concept of a relation (see Weinberg, 1965). This ontology forces one to adopt a reductive strategy of relations, since all properties, including relational ones, have to fit this ontological framework.

Bertrand Russell, among others, has effectively refuted reductive strategies. I will recall some of the discussion here, because an adequate concept of relation is in my view a *sine qua non* for a theory of artifacts. Russell distinguished two reductive theories: the monadistic and the monistic theory.[6] He described them (1938, p. 221) as follows:

> [g]iven, say, the proposition *aRb*, where *R* is some relation, the monadistic will analyze this into two propositions, which we may call ar_1 and br_2, which give to *a* and *b* respectively

adjectives supposed to be together equivalent to *R*. The monistic view, on the contrary, regards the relation as a property of the whole composed of *a* and *b*, is thus equivalent to a proposition which we may denote by *(ab)r*.

The monadistic theory has difficulties accounting for asymmetrical relations such as *greater than*. I will use Russell's example of this simple relation here, assuming that the outcome will also apply to more complex relational properties like functions. According to the monadistic theory, there must be *intrinsic* properties that explain why the relation of "greater than" is asymmetrical. Take, for example, two lines *L* and *M* which are identical in every respect, except that *L* is greater than *M*. This will be analyzed as *L is (greater than M)*. The adjective is complex, consisting of several parts. The part *greater than* by itself does not discriminate between the terms, since *M* will also have this property in some respect. The asymmetry must come therefore from the inclusion of *M* in the predicate *greater than M*. But on analysis this turns out to be a relation in a disguised form, since *greater than M* is an extrinsic predicate instead of an intrinsic one. The original relation is transformed in a relation between extrinsic predicates. The monadistic theory thus fails.

The monistic theory is equally flawed. It holds that every relational proposition *aRb* is to be resolved into a proposition concerning the whole which *a* and *b* compose. The proposition *a is greater than b* is analyzed, say, as the proposition *(ab) contains diversity of magnitude*. Asymmetrical relations are a source of trouble here again. *(ab)* is symmetrical with regard to *a* and *b*, and thus the property will be exactly the same in the case where *a* is greater than *b* as in the case where *b* is greater than *a*. "In order to distinguish a whole *(ab)* from a whole *(ba)*, as we must do if we are to explain asymmetry, we shall be forced back from the whole to the parts and their relation" (Russell, 1938, p. 225). The monistic theory is thus unable to explain what Russell called the *sense* of a relation.

These and other arguments have shown that relations cannot be reduced to intrinsic properties. They belong to an ontological category of their own. This implies that artifacts have indeed several *kinds* of properties. They have intrinsic, context-dependent, and relational properties. The resulting picture is that of technical artifacts as highly relational entities with respect to persons (designers, operators, users, etc.), time, place, and context. *There is no such thing as an artifact in itself.* What an artifact is, is determined in important respects by its relational properties.

The picture of technical artifacts as highly relational entities is similar to the one Louis Bucciarelli develops in his *Designing Engineers* (1994) from a different perspective. He develops his account mainly from a sociological and anthropological analysis of engineering design. In his view, artifacts are the result of negotiation processes in design, where each of the parties involved have their own perspective on what the artifact really is. They conceptualize this artifact from their own "object world", i.e., their own knowledge of objects, cognitive interests, background skills, etc. There is no overarching perspective that includes all others. Artifacts are always seen from a *particular* point of view, and whether or not designers talk about the *same* artifact is ultimately a pragmatic question that can only be answered in a given context.

The idea of artifacts as relational entities places them in the same category as social objects, such as laws or passports. These entities are also relational and have their function only given a particular context of human action and intentionality. An important difference, however, is that the physical realization of social artifacts is to a large extent arbitrary. Passports or identity cards can be made of any material, provided that they are not easily copied. The functions of technical artifacts such as bridges, computers, or cars, however, are partly *explained* by their physical structure.

An obvious exception seems to be software. Here we have a technical artifact the physical realization of which does not seem to be important. The increasing number of products like software is even referred to as the "dematerialization" of technology. I believe, however, that there is a misunderstanding involved here. Software as such does not have any functionality. It is just a listing of algorithms. Only when it is implemented in hardware does it get the functionality it is designed for. Or, put differently, only the *combination* of hardware and software gives us a complete artifact. This is also evident from the fact that software is always developed for a *particular* hardware context. Not any old computer will do. In my view, the right approach to this problem is to regard both hardware and software as *incomplete* artifacts. They need to be supplemented by each other to realize their functionality.

Two other interesting consequences follow from the relational character of artifacts. First, artifacts do not have clear boundaries. This is a consequence of the fact that functions have relata that often cannot be demarcated easily. As we have seen, functions supervene on both the object's physical structure and on the practice of design and use. But these practices are hard to define. What type of human action belongs to this practice and what does not? Who belongs to this practice and who does not? This is, of course, a general problem related to the idea of global supervenience. For our present discussion it suffices to note that the relational properties of artifacts makes it hard to define their boundaries: where do they begin and where do they end? A second consequence is that there is no simple answer to the question: "How many parts does an artifact have?" If functions are relative to designers and users, it means that the functional composition of an artifact may vary, depending on the perspective taken. Complex artifacts such as airplanes are divided up differently if viewed from the point of view of mechanical design, aeronautical design, electrical design, production, assembly, use, maintenance, disassembly or recycling. Each perspective will come up with a different list of parts.[7]

4. THE EMPIRICAL TURN AND THE ROLE OF CASE STUDIES

My discussion so far underlines the main objective of the empirical turn in the philosophy of technology. As we have seen, artifacts have functions, i.e., relational properties that relate objects to practices of design and use. These practices have to be taken into account in order to develop a full understanding of the nature of technical artifacts. Particularly the practice of engineering and engineering design requires special attention, given our current understanding of this practice. Methodological, epistemological and ontological issues have to be addressed here, and not just ethical or cultural ones.

The necessity to include the practices of design and use in the analysis of artifacts in itself does not yet constitute an argument for *empirical* research and case studies in the philosophy of technology. If adequate conceptualizations of these practices would exist, they could be used for the analysis of artifacts. But this is precisely what is lacking. Whereas scientific practices have been given a lot of attention in the philosophy of science recently, the situation in the philosophy of technology is very different. Here we are still in need of detailed studies of design and development.

The empirical turn in the philosophy of technology also raises the question what exactly the role of empirical data and case studies in philosophical research should be. This depends to a large extent on the conception of philosophy one has and how one conceives the relation between philosophy and the sciences. In general, the relationship between philosophy and science may be conceived in the following ways.

(1) Philosophy as proto-science. This conception builds on the fact that many sciences have emerged from philosophy (physics, biology, psychology, etc.), and that philosophy played an important role in the early development of these sciences.

(2) Philosophy as the clarification of scientific practices, methods and concepts. This view of philosophy may refer to a Wittgensteinian, therapeutic type of activity, where the fly is shown a way-out of the bottle by analyzing concepts in the practice of their use. It may also refer to a reflection on the main theoretical concepts and methods in certain scientific domains. As

a matter of fact, some sciences include such activity in their own practice, as in theoretical physics. Furthermore, the philosophy of science or the philosophies of the various sciences (physics, mathematics, biology, psychology, etc.) also aim at such a clarification. The technical sciences, however, usually contain very little theoretical reflection on their basic notions: theoretical civil engineering or theoretical mechanical engineering are not established disciplines. On the contrary, many regard the very expression "theoretical engineering" as a *contradictio in terminis*. One would expect, then, the philosophy of technology to focus on the more theoretical and conceptual issues of the technical sciences. Surprisingly enough, however, these topics are thus far largely ignored in the philosophy of science and the philosophy of technology.

(3) Philosophy as the synthesis of scientific images of the world. Here the idea is that whereas the sciences develop knowledge in increasingly limited domains, philosophy attempts to develop a synthesis of these specialized disciplines and address conflicting outcomes or interesting relations between fields.

(4) Philosophy as the clarification of the relation between the scientific and the manifest or everyday image of the world. This idea is somewhat similar to the previous one and conceives philosophy as the activity of comparing, possibly integrating and studying the interaction between types of knowledge, in this case scientific and everyday knowledge of the world.

(5) Philosophy as the critical study of science. This type of activity concerns the normative evaluation of the sciences: the way they study their object, the value of their results and the use that is made from these results.

The empirical turn in the philosophy of technology intends to refocus the discipline on topics outlined in the second approach. Clarifying basic concepts, (design) methods, and practices in the technical sciences and in the practice of engineers is necessary in this view and complementary to the currently dominant critical approach in the philosophy of technology. The empirical approach will include not only the epistemology but also the ontology of the technical sciences: the nature of technical artifacts and the way this nature is constituted in design and use by the intentionality of designers and users. The focus on clarifying basic concepts and practices in technology is different from, for example, a more sociologically oriented study of technology. Questions such as "What are basic properties of artifacts?" or "What is a design"? or "What counts as the justification of a design"? or "Is there a logic of technical functions?" and "What is an operational principle?" are usually not the type of questions the sociology or history of technology is interested in.

At the same time, the empirical turn advocates a more prominent role for empirical data and case studies in the philosophy of technology. Philosophical reflection should be based on empirically adequate descriptions of technological practices and technical artifacts. The rationale behind this is the methodological observation that the analysis of basic concepts and categories of engineering cannot be carried out in isolation from the actual practices in which these concepts are implicitly or explicitly used. The requirement to include case studies in philosophical research leads immediately, however, to the penetrating question: What would count as an adequate case of the design and use of artifacts? Do we have criteria to evaluate cases on their usefulness?

A special problem in answering this question is that artifacts are thoroughly relational objects, as we have seen in our earlier analysis. As such, they do not have clear boundaries. In addition, it can be argued that artifacts are usually part of a *system* of artifacts and a whole pattern of human activity. A train, for example, is part of a system of trains, tracks, railroad stations, electricity lines, signal and control devices, ticket offices, buildings, etc. It is also part of a system of human activities, such as the activities of engineers, conductors, drivers, passengers, cleaners, salespersons, etc. Between the two are intimate connections: technical functions are realized *only* given appropriate human actions and appropriate hardware. So how to select an appropriate case study?

In general, the following considerations are relevant, in my view, for evaluating the usefulness of cases. Most important of all, a case should be able to contribute to the clarification of basic concepts, (design) methods and practices in the engineering sciences. I call this its *heuristic value*. A case must have a good potential to discover or learn something, given a particular type of question in which one is interested. A case is not just an illustration of an already established philosophical point, but it must provide a genuine opportunity to refine philosophical analyses and concepts and to find answers to particular questions.

Secondly and related to this, a case must be rich, i.e., full of relevant details. It must contain a number of distinct features of engineering or, alternatively, of the use of technical artifacts. Such a richness of details makes it possible to encounter unexpected relations and develop new insights, which in turn may adjust the basic framework. Richness is also required for the role cases play in the "confirmation" of philosophical insights. In a sense they play a Wittgensteinian therapeutic role in that they relate conceptual frameworks developed in philosophy to the context of their use.

Thirdly, a case need not be representative in the sense that it concerns practices that are among the most widespread in the design and use of artifacts and from which empirical generalizations can be made. Philosophical research in the second approach aims at developing and clarifying basic conceptual networks, not at sociological or historical results. A philosophical result would be the clarification of the notion of technical function, for example, but not insight in the spread and distribution of a particular function, say the communication function of mobile phones in the Netherlands and its determinants. The relation between theory and data is therefore different. What counts is the richness of a case, not its representativeness. At the same time, it is pointless to build philosophical analyses on cases that no one takes serious in the engineering profession. That is the extreme to be avoided. The main criterion, however, for evaluating the usefulness of a case is its potential for answering questions, discovering new aspects and generating new insights, relative to a particular theoretical framework.[8]

5. CONCLUSION

In this chapter I have analyzed the ontological structure of technical artifacts and distinguished three types of properties: intrinsic, context-dependent, and relational properties. I argued that technical artifacts have irreducible relational properties. A central part of my argument focused on the notion of technical function. Though biological theories leave much to be desired when it comes to their application to the artifactual domain, at least two (biological) notions of function seem necessary: proper function and system function. Both are relational. The relational character of artifacts implies that there is no such a thing as the artifact in itself. What an artifact is, is determined in important respects by human actions and the intentionality involved in the practice of design and use. It also implies that artifacts do not have clear boundaries and that there is no unique way of dividing them into parts.

My ontological analysis also supports the main idea behind the empirical turn in the philosophy of technology, viz. that the practice of design and use should be taken more seriously in philosophical analysis. Case studies are needed in order to develop adequate conceptualizations of these practices. The main criterion for evaluating the usefulness of cases in the context of philosophical analysis is their heuristic value.[9]

NOTES

1. I will discuss the case of software at the end of Section 3.

2. I will not discuss here the issue whether some (physical) properties are intrinsic in an absolute sense.

3. I believe that Walter Vincenti also overstates the similarity in his "A Variation-Selection Model for the Growth of Engineering Knowledge" (1990, ch. 8). Blind variation, however, means for Vincenti not random variation, but a sort of trial and error process where the outcome cannot be foreseen or predicted as far as the matter in question is concerned.

4. The legal case against Microsoft's monopoly illustrates this very well. Microsoft artifacts partly survived because of reasons having nothing to do with their specific properties.

5. See also the discussion of the research program "The Dual Nature of Technical Artifacts", in the introduction to this volume by Kroes and Meijers.

6. The principal texts for Russell's philosophical views on the irreducibility of relations are Russell (1899) and Russell (1938), "The Classification of Relations" (1899), in *The Collected Papers of Bertrand Russell*, volume ii: 1896–1899, (Eds), N. Griffin & A. C. Lewis, London 1990, 138–146; and *The Principles of Mathematics*, especially the chapters "Relations" (pp. 95–100), "Whole and Part" (pp. 137–142), "Asymmetrical Relations" (pp. 218–226), and "Logical Arguments against Points" (pp. 445–455). For an account of the development of Russell's views on relations see Griffin (1991).

7. See also Simons and Dement (1996), who argues that even the list of materials of which an artifact consists is relative to the point of view taken: production, use, maintenance, disassembly, or recycling.

8. A beautiful example of the use of case studies in the context of an epistemological analysis is Vincenti (1990). Based on five cases in aeronautics, Vincenti develops a more general account of engineering design knowledge in his ch. 7: "The Anatomy of Engineering Design Knowledge."

9. I would like to thank Peter Kroes and Wybo Houkes for stimulating discussions on the topics of this chapter.

REFERENCES

Bucciarelli, L. L. (1994). *Designing Engineers*. Cambridge, MA: Cambridge University Press.

Cummins, R. (1975). Functional Analysis. *Journal of Philosophy*, *72*, 741–765.

Dipert, R. R. (1995). Some Issues in the Theory of Artifacts: Defining 'Artifact' and Related Notions. *Monist*, *78*, 119–135.

Dipert, R. R. (1993). *Artifacts, Art Works and Agency*. Philadelphia: Temple University Press.

Griffin, N., & Lewis, A. C. (Eds). (1990). *The Collected Papers of Bertrand Russell*, vol. 2: *Philosophical Papers, 1896–1899*. London: Unwin.

Griffin, N. (1991). Relation: The End of Russell's Apprenticeship. In: *Russell's Idealist Apprenticeship*. New York: Oxford University Press, ch. 8, 314–369.

Millikan, R. (1984). *Language, Thought and Other Biological Categories*. Cambridge (Mass.): MIT Press.

NWO Research Program. (1999). *The Dual Nature of Technical Artifacts*. (Kroes, P., Meijers, A., Franssen, M., Houkes, W., Vermaas, P.). Delft University of Technology.

Preston, B. (1998). Why Is a Wing like a Spoon? A Pluralist Theory of Function. *Journal of Philosophy*, *95*, 215–254.

Russell, B. (1938) [1903]. *The Principles of Mathematics*. New York: Norton.

Simons, P. M., & Dement, C. W. (1996). Aspects of the Mereology of Artifacts. In: R. Poli, & P. Simons (Eds), *Formal Ontology* (pp. 255–276). Dordrecht: Kluwer.

Vincenti, W. G. (1990). *What Engineers Know and How they Know It: Analytical Studies from Aeronautical History*. Baltimore: Johns Hopkins University Press.

Weinberg, J. R. (1965). The Concept of Relation: Some Observations on Its History. In: *Abstraction, Relation and Induction* (pp. 61–119). Madison: University of Wisconsin Press.

Wouters, A. G. (1999). *Explanation Without a Cause*. Doctoral thesis, University of Utrecht.

PART III

EPISTEMOLOGY AND THE EMPIRICAL TURN

THE THING-Y-NESS OF THINGS: MATERIALITY AND SPECTROCHEMICAL INSTRUMENTATION, 1937–1955

Davis Baird

1. MATTER AND IDEAS

I argue here for the importance of "materiality" to philosophers and historians of technology. My contrast is ideas. Philosophers and historians of technology should attend to the "thing-y-ness" of technological things: The fact that they occupy space; have mass; are made of materials; are subject to dust, vibration, heating and cooling; are built in "real" time and produce their work in real time; must be used – safely – by humans.

The bulk of my study concerns the development of spectrographic instrumentation for quality control use in metals production. Metal industries were transformed during the 1930s, 1940s and 1950s by the development of new methods of analyzing the metal being produced. Prior to the 1940s, quality control analysis was done "posthumously" – that is, after the melt had been poured – by wet-chemical methods that took several hours to several days to perform. By the 1950s, spectrographic methods reduced the time required to a matter of minutes, fast enough for the melt to be altered; a scrapped heat became all-but a thing of the past.

My concerns with materiality go to the heart of the philosophy and history of technology and, consequently, there are multiple respects in which these

The Empirical Turn in the Philosophy of Technology, Volume 20, pages 99–117.

ISBN: 0-7623-0755-2

concerns impact the philosophy and history of technology. For this reason, my work has a rather deep front porch. The history will come, but only after I consider the variety of reasons we should be thinking about the thing-y-ness of things.

2. BUILDING BRIDGES

With a civil engineer from the Engineering School at the University of South Carolina I co-teach a course in the philosophy of technology. Our first assignment is to build a bridge from Popsicle sticks and glue. It must span thirty inches, provide for a five-inch roadbed and hold at least one pound in the center.

For most of our students, who are not engineering majors – and even with those that are – the time spent on design and analysis is much less than that spent on construction. The major problems which students confront building these bridges are material: How can one glue the Popsicle sticks together in a straight line? How can one cope with the individual differences and irregularities in the Popsicle sticks? How should one cope with unexpected difficulties – e.g., when the sticks are not absolutely straight, or when two (or more) structural members are trying to occupy the same space? Building takes time, patience, and some faith that one's design will work, even when one has less than ideal reason to think it will. In short, the assignment reminds the students that technology is about making things, and making things is different from thinking things.

3. JOY IN MAKING STUFF

There is one other, not uncommon, result from this assignment. Generally speaking the students who are not engineering majors enjoy it. They may complain during the two weeks or so of gluing and waiting. But in the end they enjoy it. One student confessed that the sole reason he signed up for the course was that he had heard about the bridge assignment.

In his book, *What Painting Is*, artist and art historian James Elkins writes (1999, p. 6):

> But I know how strong the attraction of paint can be, and how wrong people are who assume painters merely put up with paint as a way to make pictures. I was a painter before I trained to be an art historian, and I know from experience how utterly hypnotic the act of painting can be, and how completely it can overwhelm the mind with its smells and colors and by the rhythmic motions of the brush. Having felt that, I knew something was wrong with the delicate dry erudition of art history, but for several years I wasn't sure how to fit words to those memories.

Elkins reminds us that there is a genuine human joy in making stuff. But little is written on this. Elkins notes that of the more than 8,900 books cataloged in the Library of Congress on the history, criticism and techniques of painting, less than six "address paint itself, and try to explain why it has such a powerful attraction *before* it is trained to mimic some object, *before* the painting is framed, hung, sold, exhibited and interpreted" (1999, p. 6, emphasis in the original).

My father, Walter Baird, co-founded Baird Associates in 1936. Much of what follows concerns Baird Associates' contributions to spectrochemical instrumentation. In 1940 my father wrote in his diary (Walter Baird diary entry for January 20, 1940):

> Our first spectrograph now has a new home. I am not too well pleased with its new owners for I am sure they will not treat Specky with the proper degree of affection. I could hardly expect them to. To me, that instrument represents nearly a week's work without sleep. It also represents the feeling that went into this business – something which money cannot buy. We sold it because we need money and Specky represented most of our capital. We also sold it knowing we could replace it with a more perfect instrument. It was Specky's imperfections which endeared "her" to us for I know every inch, every screw. We may build many an instrument but that one has a soul where all the others have only bodies.

My father was immersed in the materiality of Specky, every inch and every screw. He spent a week without sleep making Specky; while this work certainly involved ideas, the product was material and what was time-consuming – indeed, simply consuming – was making the material thing. He told the *Johns Hopkins Magazine* that one of the most important things he learned while a graduate student at Hopkins was "how to work with his hands – running a lathe, learning how to build something from scratch" ("RL" 1958, p. 11). While traveling on business, he wrote John Sterner, co-founder of Baird Associates, "I want to get back. I have not loafed but I want to get my hands on things again" (1936).

4. WORRIES ABOUT THE "DESIGN PARADIGM"

Practitioners spend their lives engaged with the material world. How strange then that the history and philosophy of technology focuses on ideas, not things. Walter Vincenti's terrific book, *What Engineers Know and How They Know It,* provides a telling case in point (1990, p. 10, emphasis added): "I concentrated on *ideas rather than artifact*s and sought to trace the flow of information". The irony is that Vincenti documents the intimately interwoven nature of ideas and materials, through the building and gaining experience with material models and prototypes.

Peter Kroes argues in a 1996 paper that "design" is the appropriate locus for a philosophy of technology (p. 44, emphasis in the original): "The overall assumption underlying our analysis is that an artifact acquires its concrete shape during the design process and that therefore the design process is the appropriate place to study the influence of various factors on changes in artifacts." Now, in fact, an artifact acquires its concrete shape during the making process, as a result of direct engagement with materials. I have no beef with design, which certainly is central to technological progress. However, design can easily be misunderstood exclusively in terms of ideas, and it is the exclusion of materiality that concerns me about the "design paradigm."

Focusing on ideas produces odd category mistakes. Kroes writes, for example,"Physical constraints are factual statements about nature" (1996, p. 63). I would have thought that a "physical constraint" would have to be physical, not a statement. But Kroes has already made clear that his realm of analysis is ideas, not objects; "physical constraints" are a class of ideas we have about the behavior of objects. He continues,

> many of them [physical constraints] find their origin in physically necessary (lawlike) relations. They are 'brute facts' about nature that are not negotiable at all; they transcend human power (1996, p. 63).

Here I take more serious issue with Kroes.

Brute facts should be understood physically, not in terms of physically necessary lawlike relations. Seeing them in terms of laws turns our attention away from "contingent necessities." Much design is a matter of dealing with the specific contingencies that a particular design problem presents. One problem that appears in the history that follows concerns figuring out how to cram more photo-multiplier tubes into the small space available. No doubt geometric laws are involved here. But the specific constraints involved are not given by laws; they are given by the contingencies of what was trying to be accomplished in the instrument involved. When we move the level of analysis to lawlike relations we no longer see such issues, issues which, in fact, take up much design time.

Put in other terms, I think philosophers and historians of technology should resist "semantic ascent," the move from direct talk about the physical to talk about statements purporting to describe the physical: physical constraints should be physical. Unfortunately, however, for philosophers and historians semantic ascent is seductive. Philosophers and historians work with ideas, not artifacts; this is where our intuitions and sympathies are most developed. Perhaps for this reason, but certainly in collusion with this reason, ideas have received special recognition for carrying knowledge. This is also why the Popsicle stick bridge assignment teaches so much.

The design paradigm results in a secondary status for the material aspects of technology. Indeed, relative to science, technology as a whole assumes a secondary status. This surely is the source for the persistent view of technology as applied science – despite massive historical evidence to the contrary.

5. MATERIAL EPISTOMOLOGY

Writing about technology poses an epistemological problem. Classically conceived, knowledge is propositional in nature – Plato's justified true belief. It is the kind of thing that might be preserved in libraries. Technology, by contrast, involves innovations in the material realm in an essential way; new artifacts have to be created or existing artifacts have to be modified. Much of the history of technology is the history of these material moves. It is not a history of propositions getting closer and closer to the truth. I wish here to elevate our epistemology with the insight of the "1950s American gear-head" who spent his youth making his car work better.

To properly understand the epistemology of technology we need understand the artifacts of technology as contributions to knowledge. Lacking this leaves us with little to say about engineering knowledge. Is it any wonder, then, that most philosophy of technology deals with non-epistemological aspects of technology?

Dava Sobel's wonderful *Longitude* (1995) tells the story of the "un-learned" clockmaker, John Harrison, and how he, with unparalleled wood working and clock making skills, solved a major scientific problem of the age: How to find longitudinal position at sea. He built an accurate seaworthy clock.

In the revealing and tragic conclusion to *Player Piano* (1952), Kurt Vonnegut describes a scene where an excited group of people are making vending machines work. These are the same rebel revolutionaries who spent the previous day smashing machines to emancipate themselves from the alienation of a world run by machines. Life under the machines was intolerable, but still there is genuine excitement in making something work.

Humans – some humans anyway – are inspired by learning more about the world. This learning includes material performances – making clocks no one thought possible and making vending machines vend – as well as propositional learning. One standard source of tragedy is the fact that this natural source of inspiration, elevating the human spirit, frequently runs up against political or financial forces that can kill the human spirit. Herein lies the vigor and depth behind both Sobel's non-fiction and Vonnegut's fiction.

I have argued elsewhere for a neo-Popperian approach to a *material* epistemology (2000). I won't repeat myself here. One thing a material epistemology requires is an appreciation for the difference between ideas and things, hence my current work on the thing-y-ness of things.

6. ANALYSIS IN METAL PRODUCTION 1935–1955, AND A PERSONAL NOTE

Prior to 1935 what analysis was done for the purposes of quality control in metal production was done with wet chemistry. Given the length of time required, this meant that quality control was a *post hoc* activity and metal which did not measure up to specifications had to be scrapped. Spectrochemistry changed this. By the 1950s, spectrographic methods reduced the time required to a matter of minutes, fast enough for the melt to be altered (Baird, 1991, 1993, 2000).

The bits of this history that I focus on here are taken from developments in the Baird Associates line of spectrochemical instrumentation. I conclude with "spectromet," an instrument brought to market in 1955. Baird Associates (BA) built their first grating spectrograph in 1937. In 1944, Jason Saunderson, Victor Caldecourt, and Eugene Peterson working at Dow Cemical Corporation began developing a direct reading spectrometer, one that would indicate the percentage concentration of various elements directly, without need to develop and interpret photographic spectrograms. In 1947, Dow agreed to let BA manufacture and market a modified version of their direct reader commercially. In part to cope with difficulties in the operation of installed direct readers, Saunderson left Dow to work for BA in 1952. In 1954 Saunderson began work on a self-contained spectrometer that could stand on the foundry floor – spectromet. This is the year I was born. Spectromet was brought to market in 1955.

Baird Associates – subsequently Baird Atomic, subsequently Baird Corporation – continued to develop, manufacture, and market instruments for spectrochemical analysis and quality control (among other things) throughout the 1950s, 1960s and 1970s, indeed to the present day. Baird Corporation was acquired by IMO Delavale, a conglomerate, in 1987. Subsequently it was sold to Thermo-Jarrell-Ash in 1993; Jarrell Ash was one of BA's main competitors in the 1950s and 1960s.

I began studies in philosophy of science in 1977, graduating with a dissertation in the philosophy of statistics in 1981. My father received the Scientific Apparatus Manufactures Association (SAMA) lifetime achievement award in 1979; he died in 1982. In 1983, I began to think more seriously about the role of instruments in science and technology, but not until the late 1980s did it dawn on me that the company my father founded was in this business. I seriously began to look into the history of Baird Associates, acquiring early documents and photographs, in the late 1980s.

7. HENRY ROWLAND'S SCREWS

When my father realized – to his distress – that I was pursuing a doctorate in philosophy of science – and not physics – he gave me a book, *Selected Papers of Great American Physicists* (Weart, 1976). My father inscribed the book, "To Davis, Read all – but be sure to read Henry Rowland." The book contained two articles by Rowland, his 1899 American Physical Society Presidential Address, "The Highest Aim of the Physicist," and his *Encyclopedia Britannica* (9th ed.) article, "Screw." I dismissed Rowland. What could screws have to do with the epistemology of science? Rowland was R. W. Wood's predecessor in the John's Hopkins Physics Department and Wood was one of my father's teachers at Hopkins. I thought my father's interest for Rowland was simply "familial."

Only several years later, alas, after my father died, did I give up my youthful, perhaps Freudian, rejection of Rowland's screws. These two articles present a powerful, if unappreciated, epistemological lesson. In "The Highest Aim," Rowland argues that "the aims of the physicist . . . are in part purely intellectual" (Weart, 1976, p. 102). In "Screw," we have an example of one such intellectual aim. Rowland and my father were urging me to notice something in science beyond ideas – the material.

Rowland was an expert on screws: How to make them, how to test them for various imperfections and how to use them despite their imperfections. Rowland's expertise allowed him to rule diffraction gratings that were more than an order of magnitude larger and more accurate than any previously made. Rowland's gratings made possible the spectrochemical transformation of the metal industry. Spencer Weart notes that of Rowland's various scientific contributions – to electromagnetic theory, for the authoritative determination of the value of the ohm, for the variation of the specific heat of water with temperature, and for supervising work leading to the discovery of the Hall effect – his "greatest contribution to science was the construction of diffraction gratings" (p. 84). Rowland's greatest contribution was a material contribution made possible by his expertise with screws. Without these material accomplishments, spectrographs with the requisite quantitative accuracy would not have been possible.

Rowland's gratings were made at Johns Hopkins where my father earned his doctorate. Through this connection, BA could obtain Rowland gratings (at this time produced by Rowland's successor, R. W. Wood). This was one key element to the success of BA spectrographs.

Rowland understood that no screw is perfect, and that both in order to make a better screw (from a master screw) and to use a screw, one has to work with a screw's imperfections. Rowland's ruling engine for making diffraction gratings was an exemplar of construction compensating for error.

From Rowland, I take my first lesson on the thing-y-ness of things: We can imagine a perfect screw, but we cannot make a perfect screw. And if we don't recognize that all material reality departs from the ideal of perfection – and design and build to compensate for this – we won't get very far. Ideas may be perfect, but reality is more interesting.

8. MOVING THE INSTRUMENT

In the 1930s, when my father first got into the spectrograph business, grating spectrographs suitable for quantitative analysis of metals required an entire light-tight room. A track, supported on massive concrete piers to eliminate vibration, followed the "Rowland circle" (where spectrographic images were in focus), around the room (Harrison et al., 1948, p. 30). In order to photograph different parts of the spectrum, the spectroscopist had to sneak around the room with a dim red light moving plates along the track. BA had in mind a self-contained unit that could be easily operated by people who lacked expertise in spectroscopy (Sterner, 1990).

I skip the details that went into the design and construction of BA's first 1937 three-meter spectrograph. Their original instrument, minus spark-gap and supports, could be fit in a box 2 feet by 2 feet by 12 feet. It was something that could be moved and relatively easily installed in an appropriately prepared laboratory room. The instrument was much easier to transport than an entire light-tight room.

Only people with expertise in spectroscopy could use the 1937 BA spectrograph. The Baird/Dow direct reading spectrometer, first sold in 1947, could be used by people lacking this expertise. While larger than the spectrograph, the direct reader eliminated the need for developing and interpreting photographic film – spectrographs. This reduced space requirements, for no longer was a dark room necessary, and thereby increased the transportability of the instrument. Still, a special analytical laboratory was necessary to house the instrument.

Spectromet, introduced in 1955, eliminated the need for a special room. The 12-feet by 4-feet by 6-feet cabinet contained all the necessary optics and electronics for a direct determination of the concentrations of elements of interest. The instrument was built to withstand and/or compensate for the conditions found on the foundry floor; temperature and humidity control was not necessary.

While there are obvious issues concerning size reflected in this series of improvements, here I want to draw attention to transportation. I'll return to size below. My second lesson on the thing-y-ness of things is that transportation is a serious obstacle to dispersion and progress. The spectrochemical revolution required the development of an instrument that could be built and transported to the foundry. Smaller is better because smaller can be moved. This is not simply an "external" issue about technology transfer. It is about expanding the reach of technoscience. Bruno Latour writes of expanding networks of techno-scientific data production, consumption and distribution (1987, pp. 250ff). While I don't go along with Latour's notion that these networks are made of paper, I am persuaded by his concept of the expanding domain of technoscience. Transportation is central to this expansion.

9. THE IMPORTANCE OF MATERIALS

The direct reading spectrometer was developed to meet a specific World War II mandate: to improve the quality and the rate of production of magnesium used in aircraft. The basic innovation which Jason Saunderson and his co-workers at Dow Chemical developed was the substitution of photo-multiplier tubes for photographic film. The amount of light hitting the photo-multiplier tubes could be measured directly, electronically, while spectrograms on photographic film had to be developed and interpreted. Saunderson et. al. had to cope with a variety of difficulties in developing and perfecting their new instrument. Here I mention two, condensers and wire.

Condensers: Saunderson et al. needed good condensers to accumulate the charge received by the photo-multiplier tubes. Measuring this charge would then allow an electronic determination of the concentration of the element producing light at the wavelength picked up by the particular photo-multiplier tube in question. The best commercially available condensers were not good enough; they retained 0.4% of their charge between runs. Saunderson et al. developed a new condenser with a polystyrene dielectric, which turned out to be ideal for their purposes; such condensers retained only 0.02% of their charge between runs (Saunderson et al., 1945).

Wire: Eugene Peterson remembered one lesson they learned the hard way (1997):

> We made some mistakes in using what was at hand. We used some experimental insulated wire which had been made to check out a Dow plastic for such use. It was excellent wire but the copper was solid, not stranded. In those locations where parts received even slight mechanical stress, the solder joints eventually cracked. Little by little we replaced the solid wire with stranded.

Considered from the level of ideas, one is inclined to think that wire is wire – that line on the circuit diagram that connects one component with another. But because real wire is used in real situations where there are real mechanical stresses (however slight), the actual material performing the function matters a great deal. Stranded wire works reliably, solid does not.

My third lesson on the thing-y-ness of things is that there is a distance between materials conceived "ideally" and real materials. To work, Saunderson et al. needed a condenser, which did not retain charge when discharged. From the standpoint of ideas, when a condenser is discharged, its charge is removed. From the standpoint of materials, its charge is *largely* removed. The difference between success and failure of the direct reader was determined by the ability of its inventors to find a material for the condenser dielectric that reduced this distance between reality and ideality. The difference between solid and stranded wire makes the same point, perhaps in more prosaic terms.

10. FINDING SPACE

In 1947, BA negotiated a license with Dow to manufacture and market a modified version of the Saunderson direct reader. As BA worked to modify the Dow direct reader into a commercially viable instrument, a problem of space developed. One photo-multiplier tube is needed for each element to be analyzed. How could many tubes be fit in the space available? Much later, in his 1979 SAMA award acceptance speech, my father described the situation (1979, p. 14):

> The problem was to record more elements from a single exposure. We had accomplished five, six, seven. Now, how to do more?
>
> The original photo-multiplier tubes were about the size of a jar of jam. On the way back [from Dow Chemical in Midland, Michigan to BA in Cambridge, Massachusetts] we took a plane, private, to get on a train for Boston. We had Drawing Room A. My friend, H. M. O'Bryan, thought I had the money. I thought he had. No money!!
>
> We ended selling the side bench (Drawing Room A) to a chap who needed sleep for the price of the whole room.
>
> But the moral of the story. We invented the way to turn photo-multipliers upside down, doubling the number of receivers. Our record, under considerable duress, some years later, was sixty plus – about the limit.

Unable to sleep on the train, Henry O'Bryan and my father solved a space problem.

My fourth lesson on the thing-y-ness of things recalls modern philosophy's distinction between mind and matter. Matter is extended, and the manner in which material objects are extended in space – their "figure" – can pose significant constraints on design. Jam jar geometry matters.

11. FINDING TIME

The trajectory of spectrochemical instrumentation has been a steady decreasing of size, from entire light-tight rooms, to self-contained laboratories. It has also been a steady decreasing of the time necessary for analysis. Wet chemistry took too long for useful prophylactic quality control. Spectrographs cut the necessary time to under an hour, tantalizingly close to something useful in controlling the melt. Still half an hour is too long, both because keeping a pot of molten metal sitting around waiting for the lab report, is expensive, and because the constituents in the melt are still changing during the waiting period. Direct reading held the promise of useful quality control.

By bringing the time for analysis down to the order of minutes, Saunderson et al. pushed the time bottleneck outside the instrument. Now the problem was getting the sample from the melt to the instrument and getting the information from the instrument back to the foundry floor. Consequently, even before Saunderson and his team had successfully demonstrated their direct reader, Dow went ahead and built an air-conditioned analytical laboratory to house the instrument in the basement under the foundry. The lab was connected with the foundry floor by pneumatic tubes. Here was the solution to the new time problem. With the development of spectromet, the instrument could be brought right to the foundry floor, saving yet more time.

My fifth lesson on the thing-y-ness of things, parallel to my fourth, is that time matters. Ideas may or may not be timeless. Metal alloy, in its molten state, certainly is not. Time very much matters in its production.

12. SAFETY AND ERGONOMICS

In a financial analysis of BA's products written in 1958, BA Treasurer Francis Chamberlain concluded that the original 1937 grating spectrograph – still being sold in 1958 – was not profitable (p. 12):

> When the instrument was housed in a wooden case and employed an open (dangerous) electrode stand, the margin was reasonably good, but ever since it was all enclosed in steel costs have been too high to show a net profit.

I include Chamberlain's remarks here as a simple reminder of aspects of the thing-y-ness of things that I have not discussed. BA's original spectrograph was dangerous. The electrode, where samples were sparked – brought to luminescence by a high voltage differential – was dangerous. In the first place, workers could easily slip and get parts of their body sparked. In the second place, these sparks produced stray UV light, which can sunburn an operator's eyes. Eugene

Peterson recalls suffering from such eye burn while working on the direct reader at Dow. After photographing the spectra of many samples of magnesium, Peterson burned his eyes (1997):

> The effect appeared after I had gone to sleep that night, and I woke to pain in my eyes, like they were full of sand, and unable to see due to the copious tears. I was pretty scared, I can tell you, but by morning I could see again and went to work – "there was a war on".

Things, in their thing-y-ness, can be dangerous. They must be designed to be used by humans, and used safely by humans. This is my sixth lesson on the thing-y-ness of things.

13. SPECTROMET

By 1952, Baird Associates had sold 37 direct readers. Unfortunately many of the BA direct readers were not operating reliably. Payments were being held back. In part to remedy this situation, and in part to further develop this line of their business, BA induced Jason Saunderson to leave Dow Chemical and come to BA. Saunderson spent much of his first year at BA on the road, visiting each malfunctioning direct reader and putting it right.

One of the problems that he found repeatedly was misaligned optics. Part of the reason for this situation was organizational. When Saunderson arrived at BA the Engineering Department was distinct from the Service Department. Since each instrument had to be customized – each customer would have slightly different analytical needs and thus different spectrum lines needed for analysis – Engineering would ship an unfinished instrument to be installed and customized by Service. This was unsatisfactory for several reasons. In the first place, it is much harder to accomplish optical alignment in the field. In the second place, Service personnel did not have adequate training for the job. Saunderson merged Engineering and Service, with himself as head (Saunderson, 1998b):

> I set up the optical alignment facility in a special room, trained a man to do nothing but see to it that this all-important task was done perfectly. I issued an order that no one was to make any changes in optical alignment in the field, without my knowledge, under penalty of being fired. Unfortunately this order was carried out once.

By the late 1940s, technical people interested in spectrochemical analysis had jobs in several different institutional settings. Some went to work for companies that manufactured metals and metal parts, others to instrument making firms; government laboratories and academic positions attracted yet others. Saunderson, as one case in point, went from the first kind of setting to the second. People in these various settings met through conferences, symposia and business deals and long-term friendships were established. It was through one of these friendships that spectromet was born.

In 1953 George Nahstoll, an analyst at Ford Motor Company, contacted Paul Hutchinson, a salesman at BA. Ford wanted direct reading equipment on the foundry floor instead of in an elaborate laboratory. Ford wanted two instruments – one to serve as a back-up – for its Cold Heading plant where wire was turned into bolts. These instruments needed to be as accurate as the direct readers currently made by BA, but they did not have to analyze as many elements (Saunderson 1998a).

Designing a self-contained direct reader posed three primary problems: (1) space; (2) reliable electronics; (3) optical alignment. Because of the reduced number of elements for analysis, the space problem was not critical. By carefully sealing the electronics and optics, problems with dust and rapid humidity changes were solved. The major obstacle was optical alignment; temperature changes, which are to be expected on a foundry floor, have significant effects on optical alignment.

Saunderson's solution to the problem of optical alignment was automatic monitor (Saunderson, 1998b):

> Optical alignment was maintained by balancing two halves of a mercury line during the time that the instrument was idle. That is, the time just before starting to run a sample. This was done by having a servo motor rotate a 2 mm thick quartz plate placed directly behind the entrance slit. This rotation produced small lateral displacements of all the spectral lines simultaneously.

A mercury lamp, built into the instrument would provide light for a "mercury line." This line would be focused on a special mercury exit slit and photo-multiplier tube. By a clever split-slit, the instrument could sense whether the optics were moving out of alignment (Saunderson, 1998b):

> My chief assistant, Eliot DuBois, and an engineer we borrowed from Research, Bob Burleigh, did the actual work leading up to a demonstration we put on for the whole staff one afternoon. We had an optical alignment room which I had had set-up in which we did all the optical alignment on massive steel "A" frames for the DRs. This involved focusing the grating and aligning all the exit slits on the proper spectral lines. So I had the staff gathered around the "A" frames in the darkened room, and I asked someone standing alongside to lightly press down on the frame with their little finger. Of course this small pressure on the six inch I-beam frame displaced the spectrum lines, and the servo motor could be heard busily working to compensate. The demonstration was impressive . . . .

This innovation, which subsequently was added to all the direct readers, solved the problem of optical alignment, even under the harsh conditions of the foundry floor. Saunderson applied for six to eight patents to discourage the competition from copying (Saunderson et al., 1958).

Spectromet was unveiled at the "Pittsburgh Conference" in March 1955. The Pittsburgh Conference remains one of the primary annual venues for interaction between the producers and consumers of analytical instrumentation. It has grown to enormous proportions, no longer meeting in Pittsburgh – where it began

– which at least for several years did not have a conference center big enough to house the more than 12,000 participants. Spectromet was a great success.

14. BLACK BOXING THE MATERIAL WORLD

Spectromet is an example of "black boxing" technology. I use the term "black boxing" not quite in the sense common in business history where the internal workings of a business concern could be treated as a "black box" and ignored (Rosenberg, 1982). Rather I use the term closer to Bruno Latour's sense: "When many elements are made to act as one, this is what I will now call a black box" (1987, p. 131). All of the various elements which go into spectrochemical analysis – spectrum line choice, calibration, interpretation of output, etc. – are automated and built into spectromet so the instrument can operate "as one." Of "modern machines," Latour later writes that, "in the very process of their construction they disappear from sight because each part hides the other as they become darker and darker black boxes" (p. 253). Spectromet hides itself and the material world from which it provides information.

BA advertising literature notes that one "regularly trained person, probably of shop origin, may operate Spectromet" (Baird Associates, 1956, p. 3). In the language of technology studies, the instrument de-skilled the analyst's job by encapsulating all the knowledge and skill required for spectrochemical analysis in its one-button automatic operation. Another way to think about black boxing or de-skilling, is that the black boxed instrument provides the interface between the material world and the world of propositions. With spectromet an analyst or a regularly trained "person of shop origin," directly obtains propositional knowledge of a piece of the world of interest and importance, the melt: "The melt contains XYZ% cadmium; ZXW% calcium; etc." In order to accomplish this, all of the material aspects of the material world have to be transparently dealt with by the instrument.

Black boxing renders the material transparent. This is my seventh lesson on the thing-y-ness of things. When we black box we render invisible the thing-y-ness of things. In order to do this, the designers of the instrument – the black boxers – have to recognize and work with the thing-y-ness of things. All of the previous six lessons on the thing-y-ness of things matter to spectromet's design. Technical problems of size are hidden inside the cabinet. It operates in place, proving the desired information in such a manner to be directly useful to the manufacture of metal. The distance between the perfection of ideas and the reality of the material is accommodated through a careful choice of materials and compensating mechanisms such as the automatic monitor. The instrument could be easily and safely used by foundry personnel. When asked how successful the instrument was, Saunderson replied,

> Pretty successful, I would say. The performance was excellent, and the installations could be made quite rapidly (1998c).

Transportation was straightforward. Spectromet hides the materiality of steel, and the necessary material interactions to determine its qualities. The need for direct readers at Dow Chemical, during World War II came from demands on the quality of magnesium alloy used in aircraft construction. Specifically, aircraft manufacturers needed to be able to "roll out" and weld the metal easily. The calcium content in the alloy turns out to be critical to both of these functions: too much calcium and the alloy would not roll; too little and the alloy would burn when welded. Here is a material problem. It was solved by controlling for the percentage concentration of calcium in the alloy. The "informational shadow" of the molten alloy – "XYZ% calcium" – substituted for material reality. And spectromet directly provided this informational shadow.

15. MATERIALITY AND THE INFORMATION AGE

We live in the "information age." Black boxes, such as spectromet, support this notion, taking a sample of molten steel alloy as input and feeding us information. Forget the steel, what we want is its informational shadow. Analytical instruments generally have been widely viewed this way. The very first sentence of Bruce Pollard's, Richard Papp's and Larry Taylor's 1994 *Instrumental Methods for Determining Elements* is, "A request for qualitative or quantitative information about the elemental composition of a sample is common for the analytical laboratory" (p. 1). In a similar vein, the first sentence of the 1989 third edition of Strobel and Heineman's *Chemical Instrumentation: A Systematic Approach* is, "A chemical measurement begins when information is sought about a system" (p. 1). So it is no surprise that philosophers of science describe instruments in the following terms (Rothbart & Slayden, 1994, p. 29):

> Scientists naturally understand modern instruments as information processors. From this perspective many instruments function as complex systems of detecting, transforming, and processing information from an input event, typically an instrument/specimen interface, to some output event, typically a readout of information.

Elsewhere I pursue the source and meaning of such a reduction of instruments to information (1998b). A century ago James Clerk Maxwell described instruments in terms of energy – not information (1876). Maxwell wrote at the height of the "machine age," when the steam engine and the dynamo liberated us from natural sites for harnessing energy. The control of energy was the cutting edge of progress. Now the control of information is the cutting edge of progress. This makes it easy to overlook the material basis, which this information concerns.

As a counterpoint to this notion of the world, consider the following image from a dream I had some time ago. At various times in my life, I have been susceptible to violent allergic sneezing fits. In my dream I was faced with the problem of being in the middle of a sneezing fit, but only having a single Kleenex. What to do? In my dream, I solved the problem with a Xerox machine; I would make multiple Xeroxes of my one Kleenex. On awakening – in a sneezing fit – I realized that when a Kleenex is needed, information is not enough. The thing-y-ness of things matters.

16. CONCLUSION

I close with Eugene Peterson's description of the work done to develop the direct reader at Dow (1997):

> The details were conceived in conversation, sketched out, and hardware generated from sketches. Sometimes we had parts fabricated by the main machine shop, but most were fabricated in the spectroscopy laboratory shop by Caldecourt and Peterson – "File to fit".

Working with the materials was an integral part of the creation of the instrument; design on paper was not the whole story. This is important because things are not ideas. We will misunderstand technology if we do not keep the thing-y-ness of things squarely in mind. Design, conceived as a play of ideas, neglects the essential role of experience – working materials – solving the unique material contingencies that arise. Indeed, we miss entirely a draw of technology; getting one's hands on things, for some, is a powerful incentive.

Karl Popper describes two ways to approach epistemology, the subjective way where knowledge is understood as a species of belief, and the objective way where knowledge is understood in terms of the content of propositions – believed or not (1972, ch. 3). Popper argues for the objective way. While I find Popper very attractive on this point, I would not be inclined to opt for a single approach, objective or subjective. Knowledge has both faces.

When Saunderson, Peterson, and Caldecourt were developing the direct reader, their abilities to work with materials – "file to fit" – was essential to their success. Frequently this kind of "know-how" has been discussed under the heading of tacit knowledge (Polanyi, 1966). But in the end this know-how produced issue, the direct reader. The direct reader is a public product; it can be shared and used by those without Saunderson et al.'s tacit know-how. It can be detached from its makers and assume a life of its own, helping in the production of magnesium alloy and the further development of spectrochemical analysis.

The significant point I wish to emphasize here is that it is as a material piece of the world that it has these epistemic functions. And indeed, it is only through direct engagement with materials that there is any sense to Saunderson et. al.'s know-how; "file to fit" requires real files and real things to fit together. Thus, both understood objectively – as a public detached instrument – and subjectively – in terms of the interaction of hands and eyes, materials and ideas – spectrochemical instrument knowledge can only be understood with a "thing-y" dimension.

Finished installed instruments have a kind of concreteness that can be misleading. I emphasize their thing-y-ness as an antidote to the notion that design is merely a production of mind, a play of ideas detached from its material surroundings. But one can come away from my emphasis on the thing-y-ness of things with the idea that there is a single specific given thing, e.g., *the* direct reader. Larry Bucciarelli rightly has emphasized the sense in which there can be no univocal characterization for "the thing" a group of designers are developing (1994 and 2000). The mechanical engineer, focusing on mechanical concerns sees "it" one way; the thermal engineer, focusing on thermal concerns sees "it" a different way, etc. There is no single "it."

Saunderson, coming to the direct reader with a background in optics, was primarily concerned with the optical aspects of the direct reader. Caldecourt was primarily concerned with the electronics. People in the magnesium foundry were concerned with how the machine might help them avoid scrapped metal. People at BA saw a fertile product line for metal production analysis generally. The direct reader was all of these "things." Which is to say, it had aspects, properties, uses and metaphorical meanings that gave it all these different meanings. It also, and significantly, had a thing-y-ness; it was not simply an idea.

This volume concerns the empirical turn in the philosophy of technology, the fact that recent work in the philosophy of technology has paid close attention to particular developments in our "technosphere." One of the most salient lessons of the empirical turn should be the thing-y-ness of technological things. Grand notions about the dangers of "*the* Technological society" need specific embodied technologies for their credibility. Perhaps once we have attended to these particulars, we will have a more persuasive nuanced understanding of our technological world.

REFERENCES

Baird Associates. (1956). Advertising Bulletin 42: Spectromet: Direct Reading Analysis on the Plant Floor. Personal possession of the author.

Baird D. (1991). Baird Associates' Commercial Three-Meter Grating Spectrograph and the Transformation of Analytical Chemistry. *Rittenhouse, 5*, 65–80.

Baird, D. (1993). Analytical Chemistry and the 'Big' Scientific Instrumentation Revolution. *Annals of Science, 50*, 267–290.

Baird, D. (2000). Encapsulating Knowledge: The Direct Reading Spectrometer. *Foundations of Chemistry*. *2*(1), 5–46.

Baird, D. (1998b). The Reduction of Chemical Instruments to Information. Unpublished manuscript.

Baird, W. (1936). Unpublished correspondence to Baird Associates, undated. Personal possession of the author.

Baird, W. (1940). Diary. Unpublished. Personal possession of the author.

Baird, W. (1979). Acceptance Speech: 1979 Scientific Apparatus Makers Association Award. Unpublished. Personal possession of the author.

Bucciarelli, L. (1994). *Designing Engineers*. Cambridge, Massachusetts: MIT Press.

Bucciarelli, L. (2000). Object and Social Artifact in Engineering Design. *This volume*.

Chamberlain, F. (1958). Baird-Atomic, Inc.: Principal Products, Past and Present (1936–1957). Unpublished report, 5 May 1958. Personal possession of the author.

Elkins, J. (1999). *What Painting Is*. New York: Routledge.

Harrison, G. Lord Richard Loofbourow John. (1948). *Practical Spectroscopy*. New York: Prentice Hall.

Kroes, P. (1996). Technical and Contextual Constraints in Design; An Essay on Determinants of Technological Change. In: J. Perrin & D. Vinck (Eds). *COST A4 vol. 5: The Role of Design in the Shaping of Technology* (pp. 43–76).

Lakatos, I. (1978). *The Methodology of Scientific Research Programs: Philosophical Papers*. Eds. J. Worrall & G. Currieeds, vol. 1, Cambridge: Cambridge University Press.

Latour, B. (1987). *Science in Action*. Cambridge, MA: Harvard University Press.

Maxwell, J. C. (1876). General Considerations Concerning Scientific Apparatus. *Special Loan Collection of Scientific Apparatus* (pp. 1–21). London: South Kensington Museum Science Handbooks.

Peterson, E. (1997). Letter to the author, August 2, 1997.

Polanyi, M. (1966). *The Tacit Dimension*. New York: Anchor Books.

Pollard, B., Papp, R., & Taylor, L. (1994). *Instrumental Methods for Determining Elements*. New York: VCH Publishers.

Popper, K. R. (1972). *Objective Knowledge: An Evolutionary Approach*. Oxford: Clarendon Press.

"RL." ((1958). Businessman-Scientist. *The Johns Hopkins Magazine*. *9*, 11-22.

Rosenberg, N. (1982). *Inside the Black Box: Technology and Economics*. Cambridge: Cambridge University Press.

Rothbart, D., & Slayden, S. (1994). The Epistemology of a Spectrometer. *Philosophy of Science, 61*(1), 25–38.

Saunderson, J. (1998a). Letter to the author, January 23, 1998.

Saunderson, J. (1998b). Letter to the author, January 29, 1998.

Saunderson, J. (1998c). Letter to the author, February 2, 1998.

Saunderson, J. (1998d). Letter to the author, March 20, 1998.

Saunderson, J. L., Caldecourt, V. J., & Peterson, E. W. (1945). A Photoelectric Instrument for Direct Spectrochemical Analysis. *Journal of the Optical Society of America, 35*, 681–697.

Saunderson, J. L., & DuBois, E. (1958). Automatic Means for Aligning Spectroscopic Components Patent 2,837,959, June 10, 1958.

Sobel, D. (1995). *Longitude.* New York: Walker.

Sterner, J. (1990). Interview with Davis Baird, 18 May 1990.

Strobel, H., & Heineman, W. (1989). *Chemical Instrumentation: A Systematic Approach,* 3rd ed. New York: John Wiley.

Vincenti, W. (1990). *What Engineers Know and How They Know It.* Baltimore: Johns Hopkins University Press.

Vonnegut, K. (1952). *Player Piano.* New York: Dell.

Weart, S. Ed. (1976). *Selected Papers of Great American Physicists: The Bicentennial Commemorative Volume of The American Physical Society 1976.* New York: American Institute of Physics.

LANGUAGES FOR ENGINEERING DESIGN: EMPIRICAL CONSTRUCTS FOR REPRESENTING OBJECTS AND ARTICULATING PROCESSES

Clive. L. Dym and Philip Brey

ABSTRACT

Design knowledge incorporates knowledge and information about designed objects and their attributes, as well as about methods and means for undertaking the design process. Such design knowledge is articulated in several different representations or languages. This chapter presents a typology of the languages of engineering design, emphasizing the representation of designed objects and the articulation and representation of the cognitive processes of design. Design languages include verbal or textual statements, drawings and graphics, formulas, and numbers. Still other design languages follow from computational styles. The languages of design and their computer-based implementations are empirical in origin, since observation reveals that these languages are derived not from an overarching theory, but from our experience in trying to understand what we do when we talk about designed objects, articulate design processes, and teach computers how to do these things as well.

The Empirical Turn in the Philosophy of Technology, Volume 20, pages 119–148.
2000 by Elsevier Science Ltd.
ISBN: 0-7623-0755-2

In addition to presenting a typology of the languages of engineering design and discussing the role of these languages in design activity, we also discuss the possibility of automating design activity through the design and manufacture of expert systems for product design. We will use PRIDE as an exemplar of the more advanced systems of this sort in order to discuss the possibilities and limits of automating design through the use of expert systems.

1. INTRODUCTION

This study arises from a shared interest in the kinds of knowledge employed by engineers when doing design. We are interested in a better understanding of engineering design, and we believe that such an understanding should centrally involve an analysis of design knowledge and its cognitive role in the design process. We hold that the act of designing can be analyzed as a cognitive process, and more specifically as a problem-solving process, in which designers go through a series of cognitive steps to solve design problems in order to arrive at a design that meets a sets of objectives while adhering to a set of constraints. An analysis of design as a cognitive process should, we claim, identify the various sorts of knowledge used in design (which we call design knowledge), and the cognitive processes and procedures by which such knowledge is employed by designers.

It is not our purpose in this study, however, to offer a full-blown account of design knowledge or the cognitive structure of the design process. We focus instead on two more specific, interrelated issues that we believe can be helpful in elucidating both design knowledge and the cognitive structure of design processes. The first topic of our concern is the nature and the role of external representational media, or *languages*, as we will call them, in design activity. By external representational media (or representations) we mean representational forms (e.g., symbols, images) that can be found outside the human mind. They contrast with the inner mental representations that cognitive scientists have argued exist in the human mind. Engineers use various sorts of external representations in design activity, such as verbal descriptions, mathematical equations, sketches, and graphs. We want to arrive at a typology of these sundry representational forms, and assess their role in design.

The second issue concerns the possibility of automating design activity through the design and manufacture of expert systems for product design. More and more design activity relies on the use of computers, and there have been efforts to actually automate design processes through the use of so-called knowledge-based (expert) Systems (KBESes) (Mittal, Dym & Morjaria, 1986; Dym, Henchey et al., 1988; Dym, Summers et al., 1995). We will be looking at one of the most advanced systems of this sort, the PRIDE system (Mittal,

Dym & Morjaria, 1986), and use our study of PRIDE to discuss the possibilities and limits for automating design through the use of expert systems.

We would like to emphasize the ways in which these two topics are related. For us, studying external representations in design and studying expert systems for design constitute different ways in which both design knowledge and the cognitive structure of design processes can be elucidated. Moreover, they are approaches that may benefit each other. An analysis of the role of external representations in design is likely to be helpful in an assessment of possibilities and limitations of automating design through the use of expert systems, because an expert system is itself an external representational medium that embodies computational representations of design knowledge. Thus, any general considerations on the role of external representations in design *ipso facto* also apply to the computational representations found in expert systems. In other words, understanding the role of external representations in design may help us better assess the possibilities and limitations of automating design.

Conversely, we believe that our analysis of expert systems such as PRIDE can help us gain a better understanding of the role of external representations in design, because such systems are intended as models of actual design processes. We will try to assess how successful they are in this capacity, and what this implies for the role of external representations in design.

Before we embark on a discussion of these two topics, we want to explain what moves us to study design knowledge and design processes, and what we take the status and relevance of our joint project to be. We write this chapter as an engineer and a philosopher with a shared interest in engineering design. We both seek to arrive at better theoretical analyses of engineering design that describe how design processes are structured and explain how designers use their expertise to arrive at new designs. We both believe that an analysis of design knowledge and of cognitive processes and procedures in design is central to such an undertaking. We believe that the study of design requires an interdisciplinary effort. The result of our joint effort may be classified as a study in the newly emerged interdisciplinary field of design studies, to which engineering, the social sciences, and the humanities all contribute. However, we also each have our own disciplinary interests, and we would now like to describe some of these in order to explain how this research may be seen as contributing to our respective disciplines.

One of us, Dym, is an engineer, and has been motivated to study design because of a feeling, shared by many in the engineering community, that design is a misunderstood activity that is not well represented in engineering education or in engineering research. He believes that a more rigorous and scientific study of design can contribute to better design education, as well as better practice in design research (cf. Dym, 1994a). Dym, moreover, has a particular interest in artificial intelligence (AI) models of design, both as a vehicle for gaining greater understanding of the cognitive process of design, and for their practical use as design tools. Some of Dym's work is described in Sections 2 and 3 of this work wherein we propose a typology of design representations and discuss the PRIDE expert system.

The other author, Brey, is a philosopher with a research emphasis in the philosophy of technology. He is interested in studying design because it has been a neglected topic in the philosophy of technology, which has traditionally focused on the consequences of technology while neglecting the engineering practice from which technology stems. He has been concerned with the relation between design and technological impacts (Brey, 1998a, 1998b, 1999) and believes that an analytical understanding of design knowledge and design processes is required for the analysis and normative evaluation of the role of engineering design in the coming about of the societal consequences of technological innovations. Brey also believes, however, that a philosophy of design engineering constitutes a topic for philosophy in its own right that can do for engineering what the philosophy of science does for science: clarify and evaluate methods, analyze and set standards for progress, and explore ontological and epistemological issues within the discipline (Brey, 1996).

The outline of the remainder of the study is as follows. In Section 2, a typology of representational formats used in design engineering is proposed and argued for, and a role in the design process is tentatively identified for each type of representation. This section paves the way for a discussion (in Section 3) of PRIDE, an example of a successful knowledge-based (expert) System used in design. In Section 4 we address several issues that relate design knowledge and cognition to applications of artificial intelligence and their validity.

2. THE LANGUAGES OF ENGINEERING DESIGN

Design engineering, as practiced by designers, is analyzed by us as a cognitive process, and more specifically, a problem-solving process. Our analysis of design as a cognitive process is influenced in part by the successful use of programming constructs derived from artificial intelligence (AI) to elucidate design knowledge and formulate a theory of design (Dym & Levitt, 1991a; Dym, 1994a). Designers, we claim, normally find themselves working to meet a number of goals or objectives while satisfying a number of constraints in the end product of their design activity, which may be a blueprint, a list of fabrication specifications, a prototype, or some other type of design plan. Their attempt to satisfy such constraints may be analyzed as a problem-solving activity in which designers try to find design solutions that meet several objectives while satisfying multiple constraints. In this process they employ design knowledge, the specialized knowledge that defines their expertise. Much if not all of this knowledge is representational in nature: it represents artifacts (designed objects) and their attributes, as well as design processes and procedures.

In analyzing engineering design as a cognitive process, we do not mean to deny the role of noncognitive factors in design. Certainly, the inner mental processes of designers and their interaction with representational media constitute only some of the factors that determine design outcomes. We are aware, in particular, that designs are commonly explained in technology studies as largely, if not exclusively determined by social factors, including the interactions between different social groups, such as designers, corporations, government agencies, and users (Bijker, Pinch & Hughes, 1987; Bijker & Law, 1992). A complete explanation of individual designs should certainly include such factors. However, our purpose is not to explain design outcomes, but to assess the cognitive structure of design activity. We hold that sociological or economic analyses of technological innovation are well suited to explain why certain specifications came to prevail over others in a design. However, such analyses do not explain how it is that designers are able to turn a set of specifications into a design plan. We believe that this latter process can only be understood through the analysis of cognitive processes in design, even if a further explanation of these cognitive processes may again require reference to social factors.

A cognitive process is, in our analysis, not necessarily just a mental process. Certainly, design activity is made up in part by inner mental processes (e.g., thought, concept formation, and reasoning) that take place within the minds of designers. However, just as important as these mental processes is the systematic interaction of designers with various sorts of external representational media that aid them in their design activity. Designers write down their ideas on paper, using various kinds of symbolical languages, they make various kinds of graphical representations, they use computers to process information, and they take in information by consulting various sorts of information sources. An account of the cognitive structure of mathematical modeling, for example, would probably be incomplete if it would only refer to mental processes in the head of a designer and made no reference to his or her interactions with calculators, computers, or other mathematical tools that directly affect the steps taken in modeling. A proper cognitive analysis of design should therefore combine accounts of the inner mental processes of designers with accounts of their interactions with these various representational media.

In this section, we will analyze the external representational media used in design. Our aim is to present a typology of the various types of external representations used in design and to describe their role in the design process. We will call these representational media the *languages of design* because they consist of interconnected systems of representations (symbols, icons, etc.) that function like languages in which design ideas are expressed and communicated. It will become clear that external representations used by designers do not just refer to objects and their attributes. They are also used to represent design processes and procedures. We believe that this is true because designers do not just try to prescribe what a designed artifact should look like, they also continually think of the functions served by the artifact and its attributes, and of subsequent steps in their design and manufacture. For example, there is evidence that designers think about design processes when they begin to create sketches and drawings to represent the objects they are designing (Ullman, Wood & Craig, 1990).

Thus, a complete representation of designed *objects* and their attributes requires a complete representation of design *concepts* (e.g., design intentions, plans, behavior, and so on) that are more difficult to represent than are physical objects. Since the endpoint of engineering design thinking is most often a set of fabrication specifications for objects, it is important to discuss the *languages* or *representations* for both objects and processes (Dym, 1994a).

In order to describe an object, whether physical or conceptual, in detail or abstractly, there must exist a language within which that description can be written. There are several languages in which design information is cast (Dym, 1994a, 1994b):

> **Verbal** or **textual statements** are used to articulate design projects; describe objects; describe constraints or limitations, especially in design codes (see below); communicate between different members of design and manufacturing teams; and document completed designs.
>
> **Graphical representations** are used to provide pictorial descriptions of artifacts. These visual descriptions include sketches, renderings, and engineering drawings; are often interpreted within CADD systems.
>
> **Mathematical** or **analytical models** are used to express some aspect of an artifact's function or behavior, and this behavior is in turn often derived from some *physical* principle(s). Thus, this language could be viewed as a physical representation language (cf. Pahl & Beitz, 1984).
>
> **Numbers** are used to represent design information in several ways. Discrete values appear in design codes and constraints (and as attribute values, e.g., part dimensions). Numbers also appear as continuously varied parameters in design calculations or within algorithms wherein they may represent a mathematical model.

We also add to this list two text-based representations that derive from programming constructs used in symbolic computing and form core ideas of AI-based programming (Dym & Levitt, 1991a). These two representations or languages have highly stylized and specific syntaxes or grammars:

> **Rules** prescribe specific *action* to be taken in a given *situation*. Normally written in sets of IF-THEN clauses, the left-hand (IF) sides of the rules define situation(s) that must obtain before a rule is applied, while the right-hand (THEN) sides of the rules define actions to be taken. Rules are used to represent design heuristics or rules of thumb and design codes.
>
> **Objects** or **frames** are structures that can be used to represent *objects* and their *attributes*. Frames are elaborate data structures that can be linked to other frames through a variety of logic-based and/or procedural calls to other data or attributes or to calculation methods. Objects, the heart of *object-oriented programming*, are interesting because, together with their networks of links and procedures for storing and passing data, they can describe how physical and conceptual objects relate to each other.

We recognize that it is perhaps unusual to classify these two representation schemes – rooted as they are in symbolic computation – as design languages. However, *our experience has shown* that these particular constructs offer effective means for stating and applying design knowledge. We have also found that these structures can be used to organize and integrate design knowledge represented in one or more of the other design languages, thus laying the groundwork for developing integrated computational environments for design (Dym, Garrett & Rehak, 1992; Dym & Levitt, 1991b).

We could extract other constructs from symbolic programming, e.g., logic programming, neural networks, and genetic algorithms. However, we do not identify them as design languages or representations for the following reasons. *Logic programming* is really an implementation of predicate and propositional calculus that has severe limitations in terms of its expressive power. We could also argue that rules, as an extension of logic programming, serve design representation purposes quite well. *Neural networks* attempt to imitate the learning and knowledge transmission characteristics of the human brain. They function by recognizing and classifying patterns that are in turn collections of discrete numbers representing different aspects or characteristics of a problem. This representation cannot represent more structured concepts, so neural networks are limited in their ability to mimic and simulate human thought and reasoning processes. *Genetic algorithms* are used to extract local and global optima in discrete optimization problems. Here, problems are represented by strings of attributes representing potential solutions to a problem, fitness functions used to evaluate potential solutions that reflect problem constraints and objectives, and a process by which generations of potential solutions are generated by allowing reproduction and mutation of fit solutions from previous generations. Genetic algorithm representations are limited to being variations of a given topology, so their potential for representing reasoning is limited.

Clearly, we employ different languages to represent design knowledge at different times. We often cast the same knowledge in different languages in order to serve different purposes, which in turn requires that we find translations between the different languages in order to represent different aspects of an artifact or different phases of the design process. In the domain of structural engineering, for example, a complete design of a structure requires the use of many kinds of knowledge expressed in different representations (viz., Table 1). In fact, we identify experts in part by their ability to select the right language at the right time to solve a given problem. We recognize that different languages may be employed, that different representations offer different insights and utility, and that it is desirable to link these different languages in order to seamlessly model a designed artifact and the design process.

2.1. Grapical Representations

Graphical languages or representations include sketches, freehand drawings, plots and graphs, and CADD models extending from simple wire-frame drawings through elaborate solid models. Engineers have long engaged in putting "marks on paper" while doing design, and they add supporting text, lists, dimensions, and calculations to their sketches of objects and their associated functions, as well as related plots and graphs (Ullman, Wood & Craig, 1990). The support notes and other marks (and the drawings) facilitate a parallel display of information as they can be surrounded with adjacent notes, smaller pictures, formulas, and other pointers to ideas related to the object being drawn and designed. One illustrative example is a sketch made by a designer working on the packaging - to consist of a plastic envelope and the electrical contacts – to accept the batteries that provide the power for a computer clock (cf. Fig. 1). Here the designer has jotted down some manufacturing notes adjacent to the drawing of the spring contact. And, while not evident here, the designer might have scribbled modeling notes (e.g., model the spring as a cantilever of stiff-

Table 1. Alternate expressions of fundamental knowledge about plate structures.

As a *rule*:	
IF	a structural element has one dimension much thinner than the other two
AND	it is loaded in that direction
THEN	it will behave as a plate in bending
mathematically, the partial differential equation that governs the deflections of bent, thin plates is	
	$D\nabla^4 w(x, y) = q(x, y)$
verbally and *numerically*, the deflection of a floor in a residential building should not exceed its	
	length (in feet) divided by 360

ness?) or suggestions (e.g., calculate the spring stiffness as if it were a cantilever beam) or other design-related information.

Marginalia of all sorts are commonly experienced by anyone used to working in an engineering environment. Engineers doing both analysis and design often surround their pictures with text and equations. Conversely, sketches are often drawn in the margins of documents, perhaps to elaborate a verbal description, perhaps to indicate more emphatically a coordinate system or sign convention. Thus, it should come as no surprise that sketches and drawings are as essential to engineering design as any other representation.

Experience also suggests several major issues in graphical representation, as evidenced in a listing of how drawings are used in the design process. Drawings provide a permanent record of the shape or geometry of a design; facilitate the communication of ideas among designers, and between designers and manufacturing specialists; support the analysis of an evolving design; simulate the behavior or performance of a design; and ensure that a design is complete, as a picture and its associated marginalia could serve as reminders of still-undone parts of that design. In fact, experience with graphical communication is so compelling that it has been suggested that drawing is the method preferred by mechanical engineering designers to represent external data (Ullman, Wood & Craig, 1990).

Graphical representations of all kinds – whether done by hand or on a

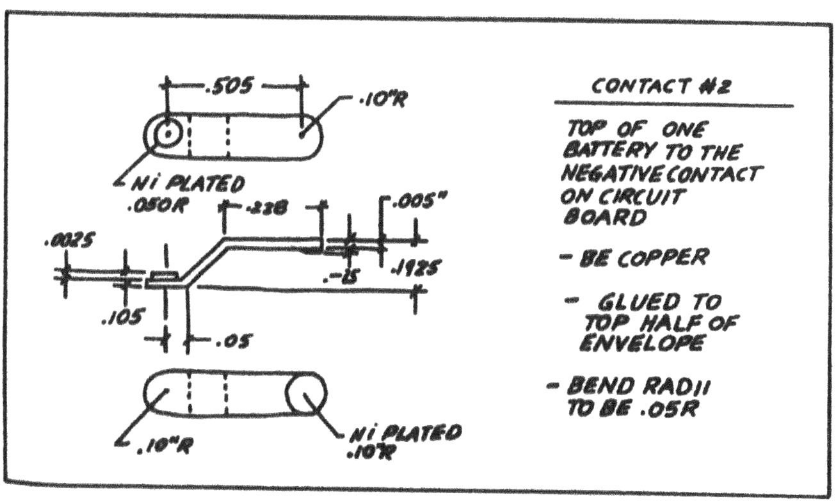

Fig. 1. Design information adjacent to a sketch of the designed object (after Ullman, Wood & Craig, 1990).

computer, whether informally sketched or detailed in complex sets of blueprints, whether bare or annotated with marginalia - represent a great deal of design information. Clearly, there is much leverage to be gained by integrating these representations with other kinds of representations (Dym & Levitt, 1991b).

2.2. Feature-Based Descriptions

We know that designers work in terms of aggregated information and concepts, for example, shapes of surfaces and volumes, holes, fit, interference, tolerances, and so on – even when they are thinking about geometry and topology. Thus, we argue that a richer, more powerful representation is needed to enhance our ability to reason about objects being drawn and designed, even if the reasoning is about spatial issues. CADD (and originally CAD) systems have been used for some time to draw pictures and plans of objects as they are being designed, but designers clearly think about artifacts in ways that are much more encompassing than the points, lines, and surfaces used in CADD systems. Feature-based representation, as it is called, has emerged from just such attempts to enable reasoning about objects for which data is stored in CADD systems (Cunningham & Dixon, 1988; Finger & Dixon, 1989).

Features were originally thought of as volumes of solids that were to be removed, typically by a machining process, because of the interest in relating a part to some part of a process plan. More recently features have been taken to include devices such as gears, bearings, and shafts, in the contexts of (1) relating an artifact's intended function to a *form feature* in the CADD representation and (2) representing the physical features of an artifact so that they can be evaluated. Examples of feature-based descriptions include: windows, corners, and tongues for injection molding; walls and fillets for extrusion; and walls and boxes for casting. Thus, the use of the term feature has broadened to include both form and function, and the context has been extended to manufacturing and life-cycle concerns of designed shapes and objects in which both geometric and behavioral issues figure prominently. This is largely due to the fact that, as noted above, designers tend to think in terms of particular forms that are intended to serve a function, imposed by the designer.

The part shown in Fig. 2 is a locator for a piece that is formed within a mold and is itself made through the process of injection molding (Dixon, Libardi & Nielsen, 1989). In order to keep the molded piece at a desired location, the shaded surfaces are required to remain parallel and at a specified distance from one another. The tongue thus serves to locate the piece within the die. Changes in any properties of the tongue must be consistent with the locating function. Thus, modifications in the tongue's geometry must be followed by corre-

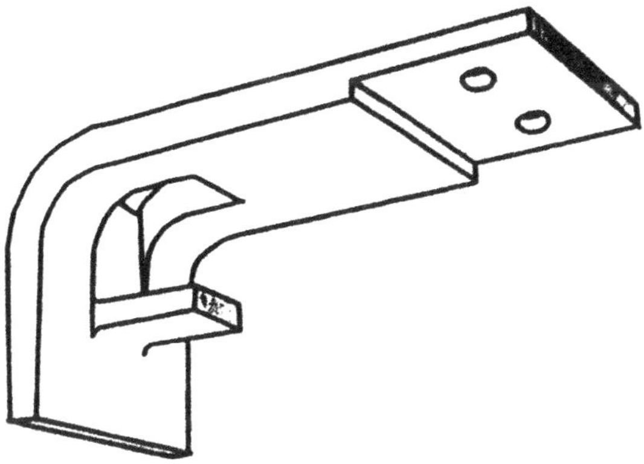

Fig. 2. An injection-molded part and some of its features (after Dixon, Libardi & Nielsen, 1989).

sponding changes in the rest of the part that do not inadvertently defeat the intended purpose. Similarly, the window over the tongue makes it easier to mold the piece being molded within a two-plate die whose parting direction is normal to the tongue, and it makes it easier to extract the part from the two-plate die. Any changes in the nature of the window must be such that they do not defeat the two purposes that the window serves in the present design.

We note that the design thinking just outlined is expressed in terms of the *tongue*, the *window*, and the purposes these features will serve. We see that this thought process is not about the coordinates that define the window opening or the precise location and orientation of the face at the end of the tongue. The thought process is done in terms of features and what features do in the final design. Experienced designers thus aggregate design information in terms of an artifact's features because they are readily available, perhaps even "intuitively obvious" choices as means to organize and apply design knowledge. Features are also a bridge between graphical representations and the object-oriented descriptions discussed in Section 2.4 because feature hierarchies are easily and naturally represented in object-oriented descriptions.

2.3. Rules

The form of a rule is well known. However, in order to show how rules are used to express design information, one brief rule example is taken from a KBES called DEEP that was built to assist designers in configuring electrical service for residential plats (Dym, Summers et al., 1995). (The design process in DEEP is described in Section 3.) The rule describes how separate building lots on a street should be clustered or grouped so that they can be efficiently configured for service. The rule combines heuristic design knowledge (i.e., a judgment about how close lots on a street are, one to another) with some basic electrical engineering knowledge (i.e., the transformer size needed to meet a specified service demand):

IF the number of lots on the street is very close or equal to the maximum number of services a 50 kVa transformer can serve
THEN cluster all of the lots on that street and serve them with a 50 kVa transformer.

2.4. Object-Oriented Descriptions

It is now appropriate to delineate object-oriented representation as a prime language of design. It clearly has evolved from the experience of both computer scientists and designers as a means of organizing or clustering information about a central core, while maintaining the ability to declare common attributes and departures from defaults, as well as to call for data or calculations that are related to the identity of the object who's central core is thus described.

The frame shown in Table 2, also taken from DEEP (Dym, Summers et al., 1995), illustrates how particular examples of a kind of object, called *instances,* can be described in relationship to a *class* of objects that share common attributes, in this case an instance of the class of transformers. This object description

Table 2. The instance XFR50_1 of the class Transformers of the DEEP system.
(There is more complexity than is visible because the links between the instance and its classes are not shown).

SLOT/ATTRIBUTE	VALUE
SuperClass	Components
Class	Transformers
InstanceName	XFR50_1
Type	BURD
DefaultType	PADMOUNT
Rating (kVa)	50
Size (ft^2)	48
Structure	Concrete
MaxConnections	6
Coordinates	(X1, Y1)
MaxCustomers	21

contains *slots* for particular attributes (the left-hand column), and the specific *values* for those attributes, that distinguish the particular transformer in question, that is, XFR50_1 (in the right-hand column). The slot for the attribute DefaultType is *inherited* or passed down from the class Transformers, here as a PADMOUNT transformer, unless it is overridden at the local level either by a rule application or through intervention by the system user. In addition, the links (not shown here) can be used to establish and maintain various kinds of relationships, for example, a CONCRETE_PAD may be *part-of* a PADMOUNT transformer, whereas a BURD is a *kind-of* transformer.

Figure 3 shows the *inheritance lattice* or object tree of all the objects in the DEEP system. The figure shows how the class Transformers and its instances (e.g., XFR50_1) fit into the overall structure of all the objects involved in this particular design project. In fact, the classes Transformers and Handholes (that are, typically, small underground structures containing electrical connection and relay devices) are themselves subordinate to a more abstract class or "super-class," Structures, which is in turn part of the top-level class that contains all the objects in the DEEP knowledge base. Note the heterogeneity of this lattice; that is, it represents a mix of different kinds of physical structures and of more conceptual objects, such as the Configurations, that represent recommendations for the spatial layout of cables and other electrical equipment.

What may also be evident from our discussion – if not from the figure, is that the branches of the object tree are not independent of one another, even though they may appear to be in the pictorial representation of the tree. For example, the transformer instance XFR50_1 has associated with it a specific location (the slot Coordinates in Table 2), and a set of lots that it serves (not shown in that figure, although the total number of lots served is given in the slot MaxCustomers). The location of this particular transformer and the lots it serves are identified in appropriate slots in objects elsewhere in the tree; that is, in the classes Configurations and Lots, respectively.

3. MODELING A DESIGN PROCESS

We now turn our discussion away from the languages of design to the modeling of design processes using knowledge-based (expert) Systems (KBESs). A KBES is a computer system that facilitates solving problems in a given field by drawing inferences from a knowledge base, which is represented in symbolical form, and which has been developed from human expertise. Work on KBESs is based on the assumption, held by many artificial intelligence researchers, that a computer, which is a universal symbol-manipulating device, has the capacity to generate intelligent action that matches that of experts in a given field

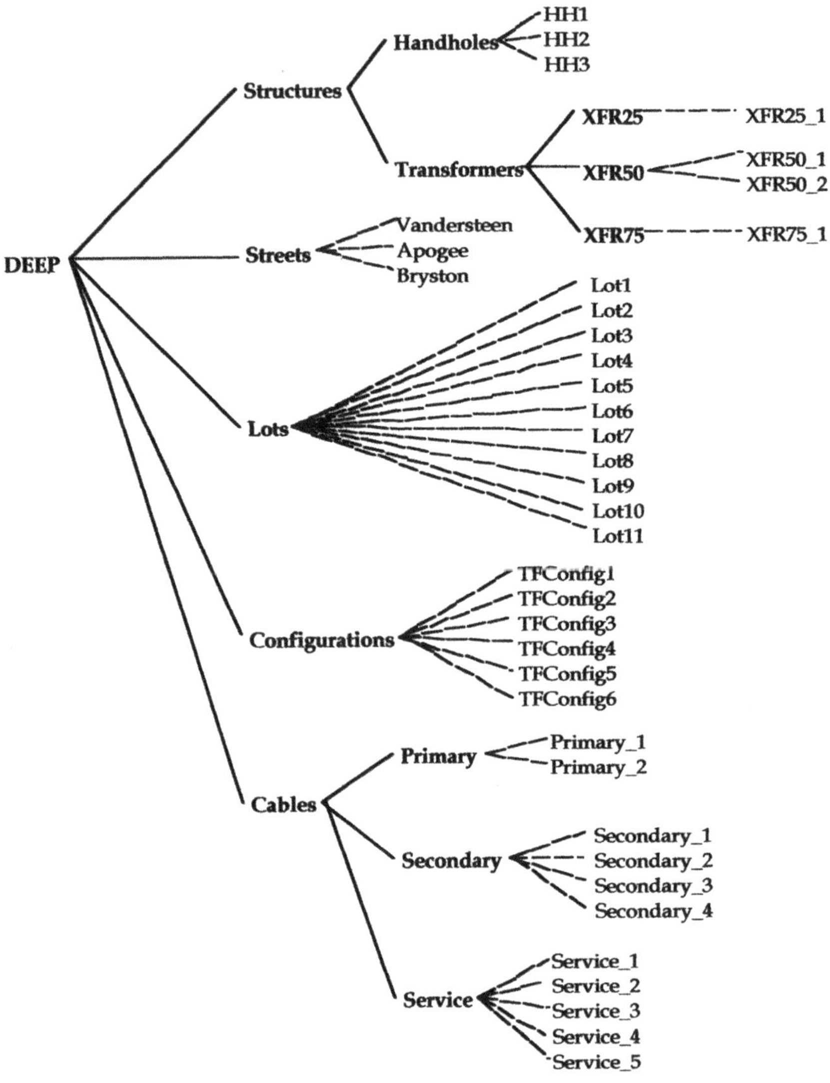

Fig. 3. The inheritance lattice (or object tree) in the DEEP KBES (Dym, Summers et al., 1995)

(cf. Newell & Simon, 1963). It is sometimes even assumed that KBESs can actually model or mimic human thought processes, because intelligence is essentially a matter of manipulating symbols according to formal rules, whether done by humans or by computers.

In this section we describe the design process encapsulated in PRIDE, a knowledge-based configuration design system for the mechanical design of paper handling systems in copier machines (Mittal, Dym & Morjaria, 1986). PRIDE is a particularly successful design system that was designed under the assumption that human design intelligence could indeed be modeled on a computer. PRIDE was also a pioneering demonstration of an empirical, AI-based approach to capturing the experiential knowledge that is applied to the performance of a design task, modeling real design processes, and furthering the understanding and vocabulary of design as a discipline (Dym & Levitt, 1991a; Dym, 1994a; Mittal & Araya, 1986; Morjaria, 1989). It is even now a "real" system, used daily as a design tool by designers doing feasibility studies for new copiers. It was not just an academic exercise, but a convincing realization of the kinds of representation and reasoning described herein. Our present discussion focuses on a few of the key points of modeling the design process.

If we looked inside a copying machine, we would see paper moving rapidly along a very complicated paper path, past various components and physical processing elements, and under rather stringent constraints. While there are several kinds of paper-handling systems, the PRIDE system focuses on transport systems that use pinch rolls to "grab" and move pieces of paper. The design requirements are tremendous and include geometrical properties (e.g., paper entrance and exit locations and angles, see Figure 4), timing constraints, allowable skew with respect to the path, tolerances on engineering parameters, and the ability to adapt to a variety of paper properties, including size, weight, stiffness, and curl. Designers are required to make qualitative and quantitative judgements while reasoning geometrically, algorithmically, and heuristically about both physical objects and more abstract concepts.

We can decompose the design of a paper transport system into subproblems, including designing a smooth path between the input and output locations, deciding the number and location of pinch-roll stations to be placed along this path, designing a "baffle" to be placed around the paper path to guide the paper, designing the sizes of various pinch rolls (drivers and idlers), selecting the proper materials for the pinch rolls and baffle, making decisions about paper travel speeds and the forces on the paper produced by the pinch rolls, calculating the time needed to move the various sizes of paper, and calculating the various performance parameters and ensuring that they satisfy the requirements. However, even with a very effective decomposition, the range of design tasks

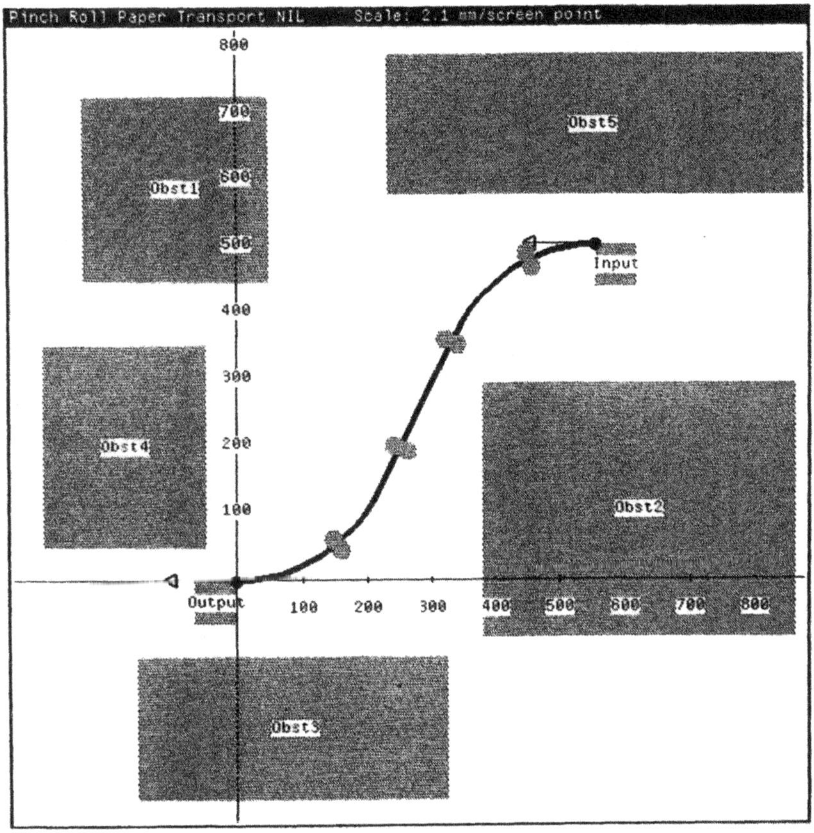

Fig. 4. A snapshot of a sample paper path in PRIDE, including roll stations, input and output points, and obstructing regions to be avoided by the transport (after Mittal, Dym & Morjaria, 1986)

we face is very diverse as it involves making decisions about geometry, spatial layout, timing, forces, jam clearance, and so on, the totality of which are often beyond the scope of a single engineer. The same transport has to be able to handle different sizes and weights of paper, which often presents conflicting constraints. For example, if the lengths (or widths) of the different sizes of paper are far apart, then the constraint on the maximum separation of neighboring roll stations for the smallest paper conflicts with the constraint on not having more than two stations guiding the paper for the longer papers. The design of the paper path is further complicated by obstructions that have to be avoided, as well as adherence to strict requirements on the smoothness, continuity, and manufacturability of the baffle in which the paper travels.

Paper-transport design is clearly a complex task that requires: *spatial reasoning* about the geometry and routing of a paper path, and the placement of particular components; *component selection* of the particular components or subsystems (from a large array) needed to perform specified functions; and *component configuration* of various subsystems in just the right way to obtain some specified behavior. Given this complexity, what does the PRIDE system do to be so helpful, either as a stand-alone design system or as a designer's assistant? In brief, PRIDE:

(1) facilitates the designer's choice of a planar path that avoids obstructions caused by equipment items within the copier and lies between specified input and exit points (see Figure 4);
(2) automatically checks that all constraints on the path geometry (e.g., smoothness, minimum radii of curvature, etc.) are satisfied;
(3) identifies a physical device (the baffle) in which the paper will be carried along the path; and
(4) identifies, designs, and locates along the path the pinch-roll pairs that grasp and move the paper along the chosen path.

In fact, PRIDE simulates rather closely the actual design process that experienced copier system designers have used for years. One of the results is that feasibility studies for preliminary copier designs are now completed and evaluated, with PRIDE's assistance, in hours rather than weeks.

PRIDE uses several representation schemes to incorporate heuristic, relational, and algorithmic aspects of the design problem, as well as several inference schemes at different levels of abstraction. PRIDE also has a powerful graphics interface that facilitates a rather complete simulation of the way human designers actually design paper-handling subsystems for copiers. Figure 5 shows a small part of an inheritance lattice that describes a paper-transport system designated as Trans5. In this object-oriented representation (cf. Section 2.4), Trans5 is an object with several attributes, some of which are linked to other objects, some of which are physical (e.g., Roll2, Driver1), and some of which are conceptual (e.g., Spec1). It can be seen that Trans5 has components that are connected to it by one or more *SubPart* links (e.g., Driver1). In order to determine which specifications govern the input point, the *InputSpecs* link is exercised to define the attributes of the design constraints at the point where the paper enters this subsystem (e.g., Specs1).

Thus, a designer can reason about the device being designed (Trans5) because questions can be asked of the system, for example, What are the output

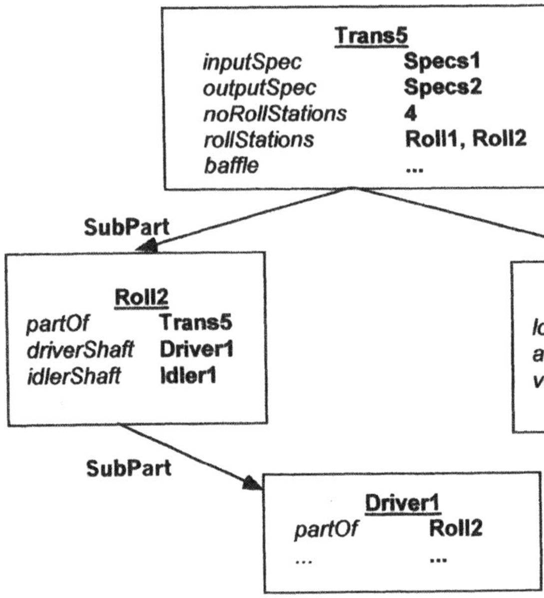

Fig. 5. A stylized version of a small portion of the inheritance lattice that makes up PRIDE's knowledge base (after Morjaria, 1989).

specifications that govern this design? and, How many and what roll stations are there in Trans5? And as is seen in Fig. 5, the answers to these two questions are, respectively, Specs2 and 4: Roll1, Roll2 . . . With this representation for devices and parts in place, attention can be turned to the design process in PRIDE.

The knowledge base in PRIDE represents a design plan structured as a top-down process of identifying and satisfying design goals and subgoals (Fig. 6). The design plan decomposes design goals into simpler steps. Fig. 6 shows a top-level goal of designing the paper transport, as well as goals for subproblems such as deciding the number of roll stations and deciding the diameter of the driver at station 4. This design plan will work only if PRIDE provides the knowledge needed to order the steps, to perform each step, to detect failures in the design requirements, and to suggest fixes for the failures.

The design process in PRIDE can be thought of as one in which a very complex system is being configured. The geometry of a path along which the paper will be transported must be established first (again, Fig. 4). After the path is established, the designers must choose components and establish their configuration (e.g., size, location, materials, etc.) so that different kinds of paper can

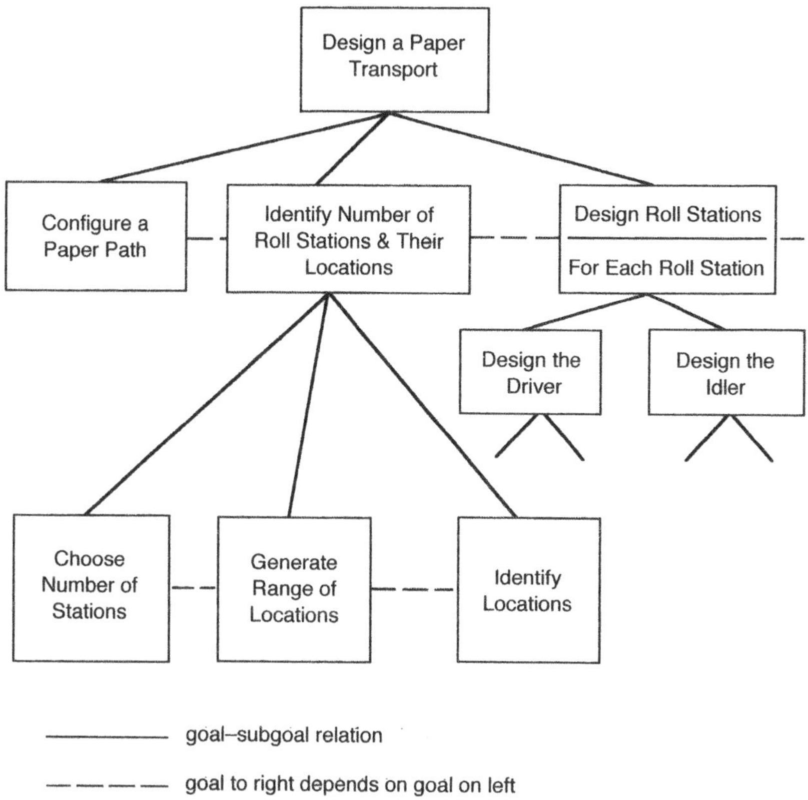

Fig. 6. A stylized version of the goals and methods that make up PRIDE's design plan (after Mittal, Dym & Morjaria, 1986; Morjaria, 1989).

be smoothly moved along the path, without jamming. The parts are chosen from a database of parts in normal use. Thus, in terms of standard design terminology, the process includes elements of both preliminary and detailed design. The process can also be characterized as *routine design* because the designers know how to decompose the design (cf. Fig. 6), how to design the subsystems and components, and what to do when a constraint is violated.

PRIDE's problem-solving strategy may be thought of as *generate-test-analyze-advise-modify*. It also makes very effective use of *decomposition* (as is evident in Figs. 5 and 6) and, although the details of this are not elaborated here, *constraint satisfaction*. A few elements of the methods by which designs are generated, and failures are analyzed and fixed, are now described.

Table 3. The goal "Decide number and location of roll stations" from the PRIDE system.

ATTRIBUTE	VALUE
Type	SimpleGoal
Name	Goal5
Descriptor	"Decide number and location of roll stations"
Status	INIT
AnteGoals	"Design Paper Path"
InputPara	"Paper Path," "length of PaperPath"
OutputPara	"Number of RollStations," "location of AllRolls"
DesignMethods	(SubGoals
	Goal51: "Decide min number of rollStns"
	Goal52: "Decide Abstract Placing"
	Goal53: "Generate Concrete Location"
	Goal54: "Build RollStn Structure")
Constraints	(Constr8: "First Stn <= 100 mm."
	Constr17: "Dist. between adj. stn <= 160mm."
	Constr24: "Dist. between adj. stn >= 50mm.")

A design goal in PRIDE is responsible for designing (and sometimes redesigning) a small set of design parameters that describe some part or aspect of the artifact being designed. Some of the design parameters in this domain are paper path segments, paper path length, number of roll stations, diameter, width, and material of each pinch roll, baffle gap, baffle material, and time taken by each size of paper during transport. We show in Table 3 a simplified representation of the goal "Decide number and location of roll stations." The variables Descriptor and Name are used to describe the goal to the (human) users of PRIDE. DesignMethods is an ordered list of all the alternate methods for achieving the goal. In this example there is only one method for carrying out this goal, that is, four subgoals be must achieved. Constraints contain the verification knowledge about the acceptability of a design.

Where does the "generate" come into the picture? In fact, one of the values of the slot DesignMethods could be a design generator, and the approach to generating designs could itself vary. It should be noted that, from the point of view of capturing a lot of design alternatives, the design generators are among the most powerful design methods. These methods are all capable of generating different values for the same (or a small set of related) design parameter(s). For example, heuristic knowledge for making "good" guesses about initial

Table 4. The generator "Design driver diameter" from the PRIDE system.

ATTRIBUTE	VALUE
type	InstanceSetGenerator
name	SetGen1
descriptor	"Generate standard driver diameters"
assignTo	(DesignObject defRollPair driver diameter)
initValue	"Find a diameter of 10mm"
classes	DriverDiameter
soFar	NIL
status	INIT

values to be generated could be attached to these methods. The generators in PRIDE also specify the ranges of possible values and increments. In Table 4 we show a design generator for the goal "Design driver diameter" that generates diameters for the drivers (in a pinch-roll pair) from a known database of acceptable driver diameters.

The particular type of generator shown in Table 4 belongs to the class InstanceSetGenerator because the database is composed of instances of different classes of objects. The generated objects are instances of DriverDiameter. It can also be seen that this method specifies that 10 mm is a good starting value for the diameter, probably because the experts have found this to be a good default choice. Finally, it specifies that this instance object becomes the value of the design parameter driver diameter.

If the current design runs into trouble, if some requirement is not satisfied, the PRIDE problem solver analyzes the current partial design and tries to come up with suggestions to overcome any violations. These modifications may be heuristics reflecting a designer's experience in fixing similar problems, or they may be based on a more general problem-solving approach which analyzes dependencies between different parts of a design to suggest modifications that go beyond knowledge directly represented in its knowledge base. Figure 7 shows how advice is provided in PRIDE. In this example, the design goal "Decide number of roll stations" calculates a number of roll stations which produces a violation of the constraint on the maximum separation between roll stations. Advice – in this case based on a built-in heuristic – is provided to say that the number of roll stations should be larger than the number calculated.

PRIDE has many features beyond those discussed here, among which is the capacity to maintain multiple designs simultaneously and to switch between

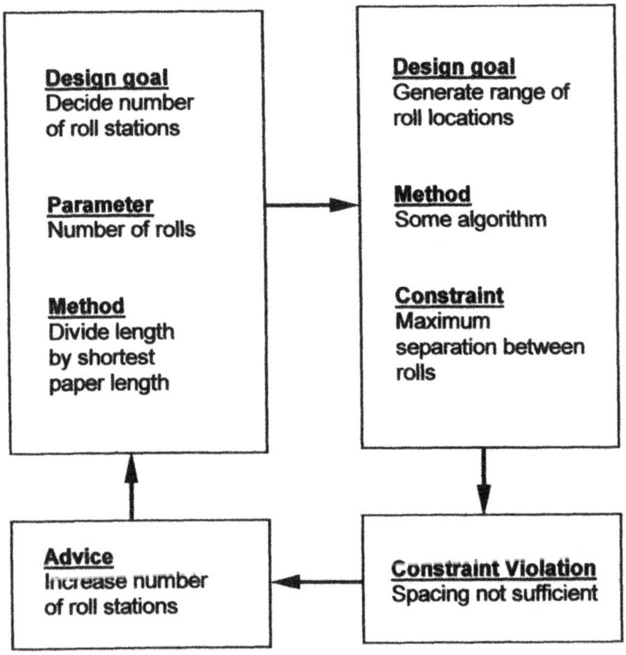

Fig. 7. A stylized view of how design advice is handled in PRIDE (after Mittal & Araya, 1986; see also Morjaria, 1989).

different partial designs so that designers can explore different options in parallel. A designer can also selectively undo a design or impose additional constraints. In fact, PRIDE's many features make it very useful as a designer's assistant, because designers working with PRIDE often develop suitable designs faster than either the system or the designer would have done alone.

Our snapshot of the design process in PRIDE illustrates how a complicated configuration task can be described and analyzed with the aid of symbolic representation and concomitant problem solving. Further, the stylized symbolic descriptions we present contrast sharply with the numerical representations used in procedural programs. This clearly opens the door to detailed and structured discussions of design, both at the blackboard and on the workstation screen, because both a vocabulary and a structure for talking about the design process can clearly be realized.

4. INTENTIONS, COGNITION, DESIGN, AND ARTIFICIAL INTELLIGENCE

We have reviewed the several languages of design, including verbal and textual statements, graphical representations and images, mathematical and analytical models, numbers, rules, and structured descriptions of objects. The last two languages in particular, derived from current computational paradigms, offer empirically based support for representing and applying various kinds of design knowledge. Inasmuch as these languages also offer the promise of capturing designers' intentions, we feel that a brief discussion of this issue is warranted.

In addition, we have postponed the question of how the languages of design function in the cognitive process of design, and specifically, how they relate to the inner mental processes of designers. And, lastly, we have postponed the question of whether PRIDE provides a realistic model of the cognitive process of design, and, more generally, of identifying the possibilities and limitations of KBESs as tools for design and design modeling. These two questions are addressed in the remainder of this section.

Designs are realizations of their designer's intentions (Dym, 1994a). However, design intentions are often subtle in their expression, or are masked by the complexity of the designed artifact. The Kansas City Hyatt Regency failure was due to the fabricator changing the structural connections for the second-floor atrium walkway because he couldn't hang it as originally designed (Pfrang & Marshall, 1982; Petroski, 1982). The fabricator found that the long hanger rods were unavailable, and he didn't know that the designer intended to hang the second-floor walkway directly from the roof truss, not from the fourth-floor walkway. Had the designer been able to convey his or her intention automatically and unambiguously, without waiting (in vain) for the question to be asked, this tragedy might well have been avoided. Thus, one wonders whether the representations described above could help convey a designer's intentions to whomever makes the final design.

Capturing the designer's intent by conjuring up a description of both the artifact and its intent is rather easy. Getting this information into the hands of the fabricators is much harder. This view of capturing design intent leaves the problem of transmitting design intent at the same abstract level as the discussion of integration. Further, it ignores some recent attempts to use design intent as a point of departure for reasoning about design. Now, the design process can be viewed as one of refining abstract goals and objectives until a fabrication specification emerges at the end of the process (Dym, 1994a). In this context, *design intent* can be viewed as a recorded history of the design process in which the *reasons* that design decisions were made are tracked as they are implemented. That is, were design intent captured systematically, design decisions could be reviewed and the consequences of revising them or undoing them altogether would be better understood. Design intent becomes a reasoning tool to help guide verification, modification, or reuse of designs, rather than simply a retrospective glance at the process.

A representation of function and its links to form and its representations is also a lively area of debate. Does the specification of function dictate a specification of form? Can one infer function from form? There is thus a need to interpret, refine, and represent at different levels of abstraction the intentions that are embodied in a design. The question is, Are issues of intention for and function of an artifact inextricably linked to the issue of representing the artifact? That is, can function be thought of independently of the representation of the device that performs that function? In fact, the examples indicate that the details of the representation of the artifact can be chosen apart from the representation of that function, subject only to the proviso that in the final analysis the designers have to produce a set of complete and unambiguous specifications for the manufacture of that artifact, which means that the representation must be acceptable to the manufacturer.

4.1. Cognition and the Languages of Design

Now to our first question: How do languages of design function in the cognitive process of design, and specifically, how do they relate to the inner mental processes of designers? We have managed thus far to resist equating cognitive processes to (inner) mental processes. Cognitive processes may take place in the mind, in which case they are mental processes, but they may also take place partially outside of the mind, as when they involve external representations and manipulations performed with those representations. For example, when someone uses pen and paper to calculate a sum, a full account of the cognitive process by which the sum is calculated should not merely describe the mental processes involved. Rather, it should also analyze the interactions that take place with this medium. After all, this medium functions as a place for inscription, storage and retrieval of information, just like human memory does. Likewise, an analysis of design activity as a cognitive process should analyze both the mental processes of designers *and* their interactions with various representational media.

At one level, we want to claim, languages of design function just like any ordinary language like English or French: they are symbolical (or iconic) forms that are decoded by designers in their effort to construct inner mental representations that are used in the cognitive process of design activity. Often, the result of this cognitive activity is one or more new external representations (e.g., a blueprint, or mathematical model). For instance, a series of textual statements that express design constraints may be read by a designer who uses them to construct a mental image of an object that satisfies these constraint, which in turn causes him to draw a graphical representation of such an object. These cognitive activities evidently require special cognitive abilities. A professional designer has interpretive abilities that enable him or her to read and understand languages of design (i.e., construct inner mental representations that denote the meaning of the external symbols or images under consideration), and cognitive abilities and background knowledge that enable him or her to translate these mental representations into new representations that move one closer to a complete design solution.

Cognitive scientist Donald Norman (1991, 1993a) has suggested that we should think of external representations as *cognitive artifacts:* tools that serve to aid and abet cognitive activity. External representations like texts and images help us in the acquisition, storage and retrieval of information. To be good cognitive tools, Norman suggests, external representations should adhere to three criteria: They (1) "capture the important critical features of the represented world while ignoring irrelevant ones," (2) "are appropriate for the person, enhancing the process of interpretation," and (3) "are appropriate for the task, enhancing the ability to make judgments, to discover relevant regularities and structures." (1993a, p. 52). These, then, are three outstanding criteria for the evaluating the use of external design representations in the design process.

We hence propose that (external) design representations should be understood as cognitive artifacts used by designers as information retrieval and storage vehicles in the cognitive process of design, for the generation of mental representations and (ultimately) novel external representations. What is needed at this point is a more detailed account of the role of external representations in cognition. This is still an ongoing research topic for cognitive science (see Norman, 1993b; Winograd & Flores, 1986; Clark, 1997; Suchman, 1987). For them to be *good* cognitive artifacts, they must be in a representational format that enables designers to encode only relevant features of a design, and that can be decoded or 'read' well by the intended user of the representation, presenting him or her with information that is, both in its form and in its content, immediately relevant to his or her goals. Evidently, the quality of design and design communication suffers when not all three of Norman's criteria are met.

4.2. Design and Artificial Intelligence

The other issue to be considered in this section is the status of KBESs as tools for design and design modeling. We are both optimistic that KBESs can play an increasingly important role in design. We are, however, divided over the question of whether they can also be understood as adequate tools for design modeling: Dym holds that they can provide good models of design processes, whereas Brey believes that this is not the case. We agree that whether KBESs can function as adequate models of the design process depends on whether they are able to adequately model the cognitive processes involved in design. This implies that there must be a structural similarity between the structure and operations of KBESs and the cognitive processes of human designers.

It is widely agreed that KBESs, like other programs run on computers, are physical symbol systems. That is, they are systems in which physically realized symbols (e.g., zeros and ones, realized as the presence or absence of an electrical current) are able to combine into symbol structures (structures composed of multiple symbols related in some physical way) and are operated on through processes of creation, modification, reproduction, and destruction. In this way, evolving collections of symbol structures are produced over time. If KBESs are to function as models or theories of design processes, it must be because the symbol structures of KBESs, and operations performed on them, mimic the representations and cognitive operations used in design by human designers. This, in turn, implies that human designers (perhaps along with the external representations used by them) are (at least by approximation) also physical symbol systems. That is, the cognitive activity of human designers must be analyzable as a process or creating, modifying, reproducing, and destroying symbol structures.

In fact, one influential account of human cognition that has been proposed in cognitive science makes exactly this assumption. This approach to cognitive science, that dates back to the very beginning of the field, features the *physical symbol systems hypothesis* proposed by Newell and Simon (1963; 1972): "A physical symbol-system has the necessary and sufficient means for general intelligent action." According to this hypothesis, intelligence is not only something that can be exhibited by a symbol-processing system like a modern digital computer. Intelligence is also essentially a matter of symbol processing according to formal rules, and therefore the human mind is *also* a physical symbol system. Mental processes are then rule-governed operations defined over symbols or strings of symbols. This model is sometimes called *cognitivism*, and when used in AI it is also called *symbolic AI.*

The assumptions of cognitivism and symbolic AI should be understood as empirical hypotheses rather than as theorems of logic or mathematics. It is worth recalling something noted by Herbert Simon (1990), one of the pioneers of both AI and the theory of design:

> AI is a field of empirical research, since it deals with systems of a complexity that does not readily yield to exact mathematical analysis. (That's one reason why synthesis is not simply reducible to analysis.)

The usefulness of KBESs as models of the design process hence depends on the plausibility of the central assumption of cognitivist cognitive science, i.e., the physical symbol system hypothesis. We do not here want to repeat all the arguments that have been proposed for and against cognitivism (e.g., Newell & Simon, 1963, 1972; Fodor & Pylyshyn, 1988; Haugeland, 1981; Dreyfus, 1992). We merely want to point out the fact that the usefulness of KBESs as models of the design process centrally depends on the truth of the physical symbol systems hypothesis, and point out that this hypothesis is still hotly debated.

We will now turn to the more practical question of how useful KBESs can be expected to be in taking over design processes. Evidently, if cognitivism is true, then one may expect future KBESs that adequately model design as it is done by humans to automate an ever-increasing part of design, with no principled limit to what can be automated. If cognitivism is false, it does not follow immediately that computers cannot be as good at design as human designers. Remember that chess programs are able to display a significant level of expertise, while they are widely believed to process information very differently than do chess grandmasters. So perhaps KBESs can be designed that follow cognitive procedures very different from human designers, but that are still successful at design.

The question is then *how successful* they may be. In answering this question, it is useful to realize that design activities encompass a spectrum that ranges from *routine* design of familiar parts and devices, through *variant* design that requires some modification in form and/or function, to truly *creative* design of new artifacts. We hold that existing KBESs, such as PRIDE, have excelled at (very complicated) routine design, and that there is evidence that they can replicate human expertise in this regard. This implies that for routine design, KBESs may not just be used as *intelligent assistants* (e.g., PRIDE), but also as automated designers (e.g., DEC's R1 system routinely configured computer systems on the factory floor, a task that had previously been performed by human sales engineers (McDermott, 1982)).

KBESs that successfully do variant and creative design have yet to be constructed. To the extent that it can guide the construction of KBESs, psychological research on expertise suggests that this would require a new generation of KBESs that are better capable of transferring knowledge to new knowledge domains and to integrate knowledge so as to create new cognitive procedures. In a review of psychological theories of expertise, cognitive scientist Keith Holyoak (1991) argues for a distinction between *routine* and *adaptive* expertise.

Routine expertise involves the ability to quickly and accurately solve problems in a limited problem domain. Experts who have routine expertise have developed increasingly specialized, skilled knowledge to solve problems in local problem domains. Adaptive expertise, however, includes the ability to transfer learned skills to new domains, and to integrate different skills so as to create new procedures out of routine expert knowledge (cf. Dörner & Schölkopf, 1991; Hatano & Inagaki, 1986). Genuine experts, and hence also, perhaps, expert systems capable of adaptive and creative design, are able to adapt their skills so as to go beyond routine expertise.

While we disagree on the usefulness of KBESs as models of design, we agree that it cannot be prejudged whether KBESs are capable of variant and creative design. The boundary between routine, variant and creative design, and between routine and adaptive expertise, is a moving one that is open to negotiation. The possibilities and limitations of KBESs for variant and creative design will reveal themselves in future generations of KBESs.

REFERENCES

Brey, P. (1996). Philosophy of Technology: A Time for Maturation. *Metascience: An International Review Journal for the History, Philosophy and Social Studies of Science, 9*, 91–104.

Brey, P. (1998a). New Media and the Quality of Life. *Society for Philosophy and Technology Quarterly, 3*, 1–23.

Brey, P. (1998b). The Politics of Computer Systems and the Ethics of Design. In: M. J. van den Hoven (Ed.), *Computer Ethics: Philosophical Enquiry*. ACM/SIGCAS Conference (pp. 64–75). Rotterdam, The Netherlands: Rotterdam University Press.

Brey, P. (1999). Design and the Social Ontology of Virtual Worlds. In: A. Nijholt, O. Donk, & B. van Dijk (Eds). *Interactions in Virtual Worlds* (pp. 5–12). Proceedings of the XVth Twente Workshop on Language Technology.

Bijker, W., & Law, J. (Eds.). (1992.) *Shaping Technology/Building Society: Studies in Sociotechnical Change*. Cambridge/London: MIT Press.

Bijker, W., Pinch, T., & Hughes, T. (Eds). (1987). *The Social Construction of Technological Systems: New Directions in the Sociology and History of Technology*. Cambridge, MA: MIT Press.

Clark, A. (1997.) *Being There: Putting Brain, Body and World Together Again*. Cambridge, MA: MIT Press.

Cunningham, J. J., & Dixon, J. R. (1988). Designing with Features: The Origin of Features. In *Proceedings of the ASME Computers in Engineering Conference*. San Francisco, CA: ASME.

Dixon, J. R., Libardi, E. C. Jr., & Nielsen, E. H. (1989). Unresolved Research Issues in Development of Design-With-Features Systems. In: M. J. Wozny, J. Turner, & K. Preiss (Eds). *Proceedings of the 1989 IFIP WG 5.2 Second Workshop on Geometric Modelling*, North-Holland, Amsterdam.

Dörner, D., & Schölkopf, J. (1991). Controlling Complex Systems: Or, Expertise as 'Grandmother's Know-How'. In: K. Ericsson & J. Smith (Eds), *Towards a General Theory of Expertise: Prospects and Limits*, Cambridge: Cambridge University Press.

Dreyfus, H. (1992). *What Computers Still Can't Do: A Critique of Artificial Reason.* Cambridge, MA: MIT Press.

Dym, C. L., Henchey, R. P., Delis, E. A., & Gonick, S. (1988). Representation and Control Issues in Automated Architectural Code Checking. *Computer-Aided Design, 20*(3), 137–145.

Dym, C. L., & Levitt, R. E. (1991a). *Knowledge-Based Systems in Engineering.* New York: McGraw-Hill.

Dym, C. L., & Levitt, R. E. (1991b). Toward the Integration of Knowledge for Engineering Modeling and Computation. *Engineering with Computers, 7*(4), 209–224.

Dym, C. L., Garrett, J. H. Jr., & Rehak, D. R. (1992). Articulating and Integrating Design Knowledge. In: *Workshop on Preliminary Stages of Engineering Analysis and Modeling.* Second International Conference on Artificial Intelligence in Design, Pittsburgh, PA, June.

Dym, C. L. (1994a). *Engineering Design: A Synthesis of Views.* New York: Cambridge University Press.

Dym, C. L. (1994b). Representing Designed Artifacts: The Languages of Engineering Design. *Archives of Computational Methods in Engineering, 1*(1), 75–108.

Dym, C. L., Summers, M. D., Demel, C. T., & Wong, C. S. (1995). DEEP: A Knowledge-Based (Expert) System for Electric Plat Design. *Computing Systems in Engineering, 6*(6), 497–509.

Finger, S., & Dixon, J. R. (1989). A Review of Research in Mechanical Engineering Design. Part II: Representations, Analysis, and Design for the Life Cycle. *Research in Engineering Design,* 1.

Fodor, J., & Pylyshyn, A. (1988). Connectionism and Cognitive Architecture: *A Critical Analysis. Cognition, 28,* 3–71.

Hatano, G., & Inagaki, K. (1986). Two Courses of Expertise. In: H. Stevenson et al., (Eds.), *Child Development and Education in Japan.* Freeman.

Haugeland, J. (1981). The Nature and Plausibility of Cognitivism. In: J. Haugeland (Ed.), *Mind Design.* Cambridge, MA: MIT Press.

Holyoak, K. (1991). Symbolic Connectionism: Toward Third Generation Theories of Expertise. In: K. Ericsson & J. Smith, (Eds.), *Towards a General Theory of Expertise: Prospects and Limits,* Cambridge: Cambridge University Press.

McDermott, J. (1982). R1: A Rule-Based Configuration of Computer Systems. *Artificial Intelligence, 19*(1), 39–88.

Mittal, S., & Araya, A. (1986). A Knowledge-Based Framework for Design. In *Proceedings of AAAI-86.* AAAI, Philadelphia, PA.

Mittal, S., Dym, C. L., & Morjaria, M. (1986). PRIDE: An Expert System for the Design of Paper Handling Systems. *IEEE Computer, 19*(7), 102–114.

Morjaria, M. (1989). Knowledge-Based Systems for Engineering Design. In *AUTOFACT '89 Conference Proceedings.* Detroit, MI.

Newell, A., & Simon, H. A. (1963). GPS: A Program that Simulates Human Thought. In: E. A. Feigenbaum & J. Feldman, (Eds.), *Computers and Thought.* New York: McGraw-Hill.

Newell, A., & Simon, H. A. (1972). *Human Problem Solving.* Englewood Cliffs, NJ: Prentice-Hall.

Norman, D. (1991). Cognitive Artifacts. In: J. M. Carroll, (Ed.), *Designing Interaction: Psychology at the Human-Computer Interface* (pp. 17–38). New York: Cambridge University Press.

Norman, D. (1993a). *Things that Make Us Smart: Defending Human Attributes in the Age of the Machine.* Reading, MA: Addison-Wesley.

Norman, D., (Ed.). (1993b). Special Issue on Situated Action. *Cognitive Science, 17*(1).

Pahl, G., & Beitz, W. (1984). *Engineering Design.* London: Design Council Books.

Petroski, H. (1982). *To Engineer Is Human.* New York, St. Martin's Press.

Pfrang, E. O., & Marshall, R. (1982). Collapse of the Kansas City Hyatt Regency Walkways. *Civil Engineering, 52*(7), 65–68.

Simon, H. A. (1990). Personal Communication, 29 October.

Suchman, L. (1987). *Plans and Situated Actions: The Problem of Human-Machine Communication.* Cambridge: Cambridge University Press.

Ullman, D. G., Wood, S., & Craig, D. (1990). The Importance of Drawing in the Mechanical Design Process. *Computers and Graphics, 14*(2), 263–274.

Winograd, T., & Flores, F. (1986). *Understanding Computers and Cognition: A New Foundation for Design.* Norwood, NJ: Ablex.

DESIGN MISTAKES: THE CASE OF THE HUBBLE SPACE TELESCOPE

Joseph C. Pitt

1. INTRODUCTION

This chapter is an exercise in what is being called the empirical turn in the philosophy of technology. As such it attempts some serious empirical work with respect to philosophical claims about technology. This may appear to suggest there is something novel here. To my ears, this latter claim presents a problem which needs to be addressed before we can proceed.

The problem is this: the effort to take an empirical turn in the philosophy of technology sounds like this is something novel and, in one sense, it is not. There has been lots of empirical work on philosophical topics dealing with technological issues. For example, one can point to Langdon Winner's investigations with respect to the electrical power industry or several of the studies on the impact of various technologies on planetary ecology. So how is this empirical turn something new? That is, how is this turn different from what has already been going on?

To answer this question, we have to understand the kind of empirical work done so far. Almost uniformly, at least in the United States, it is work with a political/ideological agenda in mind. That is, to the extent that philosophical discussions of technology have employed empirical data, it is employed in the service of an ideological perspective. The people who do this I call "the social critics". Winner (1986) is such a social critic and his work presents us with a good example of how this is accomplished. His political position, that large

The Empirical Turn in the Philosophy of Technology, Volume 20, pages 149–163.

ISBN: 0-7623-0755-2

power generating companies disenfranchise the little people because the companies, or rather their boards of governors, control the means of production, is patently Marxist. Now there is nothing wrong with having a political ideology to work from, it may even be impossible to avoid. What is wrong, and what I have been arguing for years (Pitt, 1987 and 2000), is to condemn science and technology for sins real or imagined on the basis of that or any other ideology. Or to approach it slightly differently, we make no progress in the philosophy of technology when we allow our metaphysics to frame the discussion. If we start from a blatantly metaphysical point of view, then we have eliminated the possibility of meaningful discussion, since differing metaphysical frameworks tend to be incommensurable. Now if philosophy is a dialogue in which we engage in a mutual effort to examine problems of significance, then we must be able to speak to each other, hence the methodological injunction to tend to epistemological matters first and let the metaphysics emerge as it may. In this case it will be to examine a real world situation without any particular ideological eyeglass, i.e., to find out what the facts are and what they tell us about our theories of technology.

Two objections to this position immediately emerge: (1) isn't that position itself the expression of an ideology? and (2) the facts don't speak for themselves. In the case of (1), without rehearsing the fairly detailed explanation I have developed elsewhere (Pitt, 2000), let me attempt to offer a quick denial. People use ideology to develop positions to attack other positions and to support their own political agendas. The attempt to understand, in all its complexity, the world of technology, for the sole purpose of understanding it and not to argue for or against any particular restrictions of or additional funding for any particular project is what makes the approach developed here different and non-ideological. It is the absence of a secondary objective beyond understanding that removes the ideological sting.

With respect to the second objection, more needs to be said. What we see in the world is framed by our theoretical assumptions (metaphysics?). Philosophers of science have been telling us since Gaston Bachelard (1934) and Norwood Russell Hanson (1968) that observations are theory-laden and Richard Rudner (1953) reminded us early on that scientists make value judgments, hence they would seem to employ an ideology as well, it is just in the service of truth and understanding, but an ideology none-the-less since they also argue for public support on the grounds of the merits of science. How can we expect to have the empirical turn in the philosophy of technology produce any less biased results than those of the social critics, if the observations, i.e., the facts, are also contaminated?

The difference is that here we are engaged in attempting to develop criteria for evaluating the empirical claims made about technology, which criteria ought to result in an improved ability to discern where and how ideology is employed.

In short, it is not enough to understand a technology, however contextualized. Rather, we need to know how that knowledge is constrained by the methods, assumptions, and values others or we bring to the investigation.

2. THE SCOPE OF THIS PROJECT

In that spirit, my project here looks at three different claims about the design process – WalterVincenti's, Louis Bucciarelli's, and my own – with an eye to determining how well these claims stand up in the light of real world activities. In the interest of space, only one case study is examined and even that is only partially developed, but it has the merits of having at its heart a flawed project, the Hubble Space Optical Telescope (HST), i.e., a product which embodied a design/manufacturing mistake. Here we have an example of a project gone wrong despite being eventually corrected. I argue that neither Vincenti's reiterative model nor Bucciarelli's claims about the social nature of the design process bear up when examined in light of what happened to the Hubble.

Now it should be asked why it is so important to accomplish this goal, i.e., to explain what went wrong. More is at stake than simply finding out why a project didn't meet the constraints of some theoretical model or other. It has, rather, to do with the very question of assessing the adequacy of theoretical models in general and, in turn, exposing what was referred to above as the ideological element in our models. In this case, the ideology is what I will call the "myth of the engineer", in which engineers are portrayed as the paradigms (*pace* Kuhn) of rational and project-oriented problem solvers.

In labeling this ideology the "myth of the engineer", I do not aim to offend. I seek to understand how things happen. This approach has been developing in the philosophy of science and in science and technology studies. It aims to understand how science works, Hilary Putnam's miracle, warts and all. In its infancy, the philosophy of science started with the assumption that science is our best means for achieving knowledge about the natural world and it set about attempting a justification for this claim. It just well may be true that science is our best knowledge-generating machine, but that needs to be discovered, not assumed. The fully contextualized historical and socially informed philosophy of science that is now emerging justifies the early assumption of the epistemic superiority of science, but it also exposes the mess that is real daily scientific activity and it thereby provides a deeper appreciation of the genuine accomplishments of scientists when they do emerge.

There is one more reason for looking at a project gone wrong, rather than concentrating on engineering success stories. In epistemology, one of the criteria of adequacy for any theory of knowledge is that it be able to account for how we make false knowledge claims, as well as how we come to have knowledge and how that knowledge is warranted. The process by which technological projects are accomplished is analogous in many ways to the knowledge producing process of science as well as being disanalogous in other ways (Pitt, 2000). To the extent that engineers are successful in producing products and in codifying the means for generating equal success in other projects, they develop something analogous to scientific theories and protocols. If that is correct, then philosophers of technology should be able to lay down criteria for determining if a project is a success or not. I will be attempting to show that Vincenti and Bucciarelli produce helpful models or theories which account for successful design processes, but not for ones which fail. And for those that fail we need something stronger than falling back on the position that if the project failed the participants weren't very good engineers.

I am not claiming that this is what Vincenti and Bucciarelli do, but it is what their models make possible. Attacking the professionalism of the participants shifts the blame away from the inadequacy of our models in an ad hoc manner, which is a cop-out since we seek to understand what happened, not to assess blame, which is the aim of the social critics. More to the point, if we can show that a project failed because it didn't follow our idealized models, and then turn around and explain the failure by blaming the participants for being defective in some way, we have in fact bought into "the myth of the engineer". We are in effect saying that no well-trained engineer would have allowed that to happen, and since it did happen then these folks can't be good engineers, hence this failure does not count against this particular model and its idealized account of how engineers ought to behave. To proceed in that fashion is just as flawed methodologically as Winner's claims that all big business is bad because of his Marxist orientation.

Part of what I have been arguing is that before we engage in assessing blame and social criticism, we need to know what really is or was going on. Properly done social criticism is a stage two activity. Not only does it require good detective work at stage one, but we also need to know what to look for in order to take the second step. I propose that we look at who made what decisions and why. Only by concentrating on the decision-making dimension of technological development can we proceed to responsible stage two work. (Warning: Do not confuse the talk of stage one and stage two here with the different components of my model outlined below.)

A couple of preliminary notes before proceeding to a discussion of the three models of the design process. First, I have located my discussion of technological issues in the context of engineering. There is a general problem with talking about technology per se, just as there is a problem with talking about science per se.

Generalizations about science have been shown to fail in a variety of ways, so it should come as no surprise that the same should happen to generalizations about technology. To make a long story short, talk about science needs to be talk about scientists, and particular groups of scientists and particular scientists in those groups. That is the only way to make sense of what actually happens. To accomplish similar results when discussing technology, I have tried to identify a counterpart to scientists: engineers. Like scientists, engineers have professional societies, mirroring their specialties – journals to publish in, special educational programs for credentialing, etc. I don't know what a scientist is, but I do know a biologist when I meet one. I don't know what a technologist is, but I do know what electrical engineers do. And so the following discussion of technology is contextualized within an engineering framework.

The second point concerns the focus on design. If there is a similarity to science, then the design process is to technology what the scientific method is to science. But just as there is no scientific method simpliciter, there is no design process pure and simple. There are many, and they are the methodological counterpart to methods of experimental design and testing in the sciences. In what follows I will be looking at models of the design process to determine whether or not they stand up to what engineers actually do – are they explanatory, and if so, in what sense?

3. MODELS OF THE DESIGN PROCESS

According to Vincenti (1989, p. 9):

> Design, apart from being normal or radical, is also multilevel and hierarchical. Interacting levels of design exist, depending on the nature of the immediate design task, the identity of some component of the device, or the engineering discipline required. For airplanes, which are typical devices that constitute complex systems, the levels run more or less from the top down:
>
> (1) Project definition – translation of some usually ill-defined, military or commercial technical problem for level 2.
> (2) Overall design – layout of arrangement and proportions of the airplane to meet project definition.
> (3) Major-component design – division of project into wing design, fuselage design, landing-gear design, etc.
> (4) Subdivision of areas of component design from level 3 according to engineering discipline required (e.g., aerodynamic wing design, structural wing design, mechanical wing design).
> (5) Further division of categories in level 4 into highly specific problems (aerodynamic wing design into problems of platform, airfoil section, and high-lift devices).

Such successive division resolves the airplane problem into smaller manageable subproblems, each of which can be attacked in semi-isolation. The complete design process then goes on iteratively, up and down and horizontally throughout the hierarchy.

Despite the fact that Vincenti couches his analysis in the framework of airplane design, the process seems generalizable to any other design project, be it architectural, landscaping, curriculum design, etc. Further, it allows for movement back and forth between the levels and interactive readjustment. What it leaves out, in the form in which it is expressed, are the people and the social environment in which this all takes places. Without explicitly injecting the participants into the process, the mechanism for moving between these various stages is missing. This is what Bucciarelli attempts to correct.

According to Bucciarelli (1994, p. 18):

> There is no science of design process in the way the participants understand that term. This is not to say that the process is irrational, that a story can't be developed and told that makes sense, or that one cannot, on the basis of this story, infer improvements in the process. It is to claim that to be "scientific" about the study of design process one must admit the possibility that the object – as either physical principle or economic necessity – is only part of the picture, and a very fuzzy part at that. If we want to understand the design process, we must remain sensitive to the full breath and depth of social context and historical setting.

Bucciarelli is correct to emphasize the social and the historical context and to recognize that the object is only part of the picture. Unfortunately, Bucciarelli continues a little further on to strengthen the claims about the social in ways which render his project suspect (1994, p. 20):

> My working hypothesis is that the process is not autonomous, that there is more to it than the dressing up of a scientific principle, more than the hidden-handed evolution of optimum technique to meet human needs, and more than the playing out of the bureaucratic "interests" of participants seeking power, security, or prestige. In the affirmative this hypothesis takes the form: Designing is a social process.

So far so good, Bucciarelli is taking the move to the social and distancing himself, so it seems, from the more radical social constructivists in science studies. But there is more (1994, pp. 20–21, emphasis added):

> Executive mandate, scientific law, marketplace needs – all are ingredients of the design process, but more fundamental are the norms and practices of the subculture of the firm where the object serves as icon ... In the simplest terms, design is the intersection of different object worlds. No one dictates the form of the artifact. Hence design is best seen as a social process of negotiation and consensus, a consensus somewhat awkwardly expressed in the final product.

As we shall see, the claims about the lack of outside interference on the form of the artifact are a bit too strong; likewise, the overwhelming emphasis on the norms of the firm; and it is perhaps too strong to claim that "No one dictates the form of the artifact" for, especially where government contracts are involved, there are specifications to be met. Thus while the exact form of the artifact may not be dictated, its performance measures are. I do like the emphasis on the internal workings of the firm and the notion of compromise that comes out of a realistic understanding of the negotiation process; it is a powerful antidote to armchair theorizing. Elsewhere (Pitt, 2000) I have spoken of the stark ontological primacy of compromise in the domain of the social. But the compromises take place among a large number of players; sometimes the most important ones are internal to the firm, but not always. In the case of the Hubble, as we shall see, governmental funding was a key issue that wasn't merely reducible to the internal question of the firm's profits.

In the design process, there is no one necessarily key actor who plays the major role in determining the outcome known as the product. The model I have been working with starts with the definition of technology as humanity at work and unpacks that definition in terms of a multilevel in-put/out-put transformation process. The basis for this model is Glendon Schubert's (1965) analysis of the structure of decision-making in the United Sates Supreme Court. Consistent with Schubert's view of the nature of the decision-making/policy-making process is the consequence that there are input/output transformation processes whose function is to develop other input/output transformation processes. Let us distinguish between these as first-order and second-order transformations. Decisions are first-order transformations. The result of a first-order transformation may be either another first-order transformation, i.e., a decision to make another decision, or a second-order transformation, i.e., a decision to create a tool of some sort.

A second-order transformation involves the construction of a device to meet specific goals. An oil refinery is a second-order transformation. So is a legal system or a geometry. They are the result of first-order transformations in which a decision was made using available knowledge, etc., to build, for example, a space telescope. Thus, decision-making procedures are first-order transformation processes. In the case of the construction of a space telescope we have a nice complicated example because the decision to build the telescope actually amounts to authorizing another series of first-order transformers which are the processes to be used for planning, designing, testing, and construction of the project. The building of the telescope involves further decisions as well, such as the manipulation of materials. The completed telescope is itself a second-order transformer since it transforms raw materials by, roughly, mechanical means. So, using the basic notion of an input/output process we can still distinguish between mechanical and social processes and decision-making processes, thereby allowing such institutionalized decision-making processes as bureaucracies and funding agencies to be characterized as technologies.

But to characterize technology as humanity at work, meaning by that the set of first and second-order input/output transformations, doesn't mean that we can't analyze this account further. To complete the model we need more than the notion of levels of input/output transformations. There is a crucial third ingredient we must include if our model is going to reflect the most important component of humanity at work, where that work is acknowledged to take place in a social context with a number of players both internal to the decision making team and external to it. The final ingredient is assessment feedback.

Technology assessment is a special kind of decision-making in which the effects of implementing decisions of the first kind are illuminated by means of a feedback mechanism that makes it possible to upgrade the knowledge base for further decision-making. The scientific process, as one aspect of humanity at work, may be its best general example. In a science the constant reassessment of theoretical assumptions in the light of new results, both empirical and theoretical, is essential to the development of viable theories.

But more to the point, the most important aspect of contemporary technologies is the extent to which an assessment feedback mechanism is formally incorporated into the decision-making procedures surrounding the development and implementation of plans for new ventures. In some cases this is mandated by government. But more interesting is the situation today in which the importance of assessments and feedback loops is being insisted on because of the magnitude of some technological ventures and their potential consequences. It even appears that assessment has become an important value governing humanity's work. A full-scale discussion of this phenomenon (not to be undertaken here), would reveal the means by which changing goals and values affect the development and implementation of new and innovative techniques for transforming raw materials into suitable results in both the physical and social domains. Let me just mention one such change to illustrate the point. In the Netherlands, when a new oil refinery is proposed, it must include in its plans not only the usual environmental impact statements, but a step by step process for dismantling the plant when its working life is over and for the disposal of its various components. This is taking environmental impact assessment to the next level.

At this point it should be apparent that the proposed model is intended only to schematize the complexity and pervasiveness of technology rather than to be a definite description of its structure. On any detailed analysis it will become clear that wherever a decision is involved, so too are a variety of other considerations, among them the assessment process, the nature of the second-order transformer, i.e., mechanical or social, the social circumstances, the goals of the individuals as well as the goals of the institutions, and so one cascades down a virtually unclosed spiral. The success of this model, then, should not be judged in terms of whether it simplifies our view or not; it doesn't, nor was it intended to. The world is a very complex place and we do ourselves serious disservice by thinking that we can always get around in it better by simplifying things. The merits of the proposed account are to be found in the manner in which it allows the complexity of the actual situation to be exhibited, while still providing the means to isolate and analyze the relevant components and their interaction in the light of constantly changing circumstances.

4. WHAT HAPPENED TO THE HUBBLE?

We now have our three models. Let us turn to the problem at hand. How did NASA manage to launch a defective telescope? Or, not to appear to put the blame on NASA, how did a defective telescope come to be launched? Most of the material I will use here comes from the *United States Congressional Record* (November 16, 1993). The U.S. Congress, following the discovery of the blurry images coming back from the Hubble, ordered an investigation. This resulted in a report presented to Congress by William Colvin, Inspector General of NASA. Several things are not in dispute: (1) The primary mirror of the HST was flawed. It suffered from a spherical aberration. (2) The flaw was the result of a manufacturing error and several failures on the part of management. (3) The manufacturer of the mirror, following the investigation, agreed to repay the government $25 million. The willingness of the manufacturer, Perkin-Elmer, to agree to this settlement, suggests something more than a mistake was involved.

The problem was that the main mirror of the Hubble suffered from a spherical aberration. According to Eric Chaisson (1994, p. 174):

> [S]pherical aberration is an undesired optical condition induced by a circular mirror having a deformed figure – that is, a curvature of the wrong "prescription." It has nothing to do with the smoothness of the mirror, but depends only, yet critically, on the shape of the mirror. Unable to focus all the light to a single point, such a misshapen mirror actually displays a range of foci. Alignment of the optics can emphasize the light captured either by the outer or by the inner parts of the mirror, but . . . the resulting image, regardless of these attempts, is a bright point surrounded by a disk of diffuse light.

The revelation that the telescope suffered from this specific problem and not from a failure to focus the primary and secondary mirrors was first proposed by scientists at the Space Science Institute. The NASA engineers and the engineers at the company which manufactured the mirror, Perkin-Elmer, initially objected and sought to find the solution in adjusting the two mirrors for focus.

After all efforts were exhausted, the engineers agreed to the diagnosis of a spherical aberration.

This was an expensive project, paid for by tax payers dollars – so naturally there had to be a Congressional inquiry. The result of the House Inquiry was an extensive investigation with a report from NASA following. The investigation revealed the following six significant irregularities (and here I will quote directly form the testimony of Bill Colvin):

1. **Non-approved Reflective Null Corrector Washers**. In the process of adjusting the spacings of the reflective null corrector for the HST mirror, technicians discovered they could not move the field lens into the prescribed position. Instead of calling in the designer, the contractor inserted an ordinary washer under each of the bolts holding the field lens retainer to the adjustable plate. The insertion of these non-approved washers in an instrument whose precision is measured in tens of millionths of an inch required a non-conformance report. There is no written evidence that a report was generated.

A non-conformance report is a report describing how what was done deviated from the specifications of the government design. Why did the Perkin-Elmer engineers fail to file such a report? In an article in *Science*, the explanation rendered is that officials at both NASA and Perkin-Elmer "allowed themselves to be overwhelmed by the massive cost overruns and schedule slippages in other parts of the project. As a result they neglected the mirror work, which seemed to be relatively well, and failed to enforce their own quality assurance procedures". This would seem to argue against both the Vincenti and Bucciarelli models. For not only do we not have evidence here of the reiterative process Vincenti proscribes, which should have intervened when the technicians couldn't move the mirror into the right position and should have called for the designer, but didn't; but Bucciarelli's confidence in the firm's concern with the object seems to be missing as well. My model focuses on the decision making and it can deal with this situation by concentrating on why the decision was made not to call the designer back in: cost overruns and schedule delays, which are externalities with a major impact on the final product. Vincenti emphasizes the structure of the design process. Bucciarelli stresses the social context, especially that of the firm. My emphasis is on the decision making and the factors that bear on it. In the remaining five items in Colvin's reports, we will see that each point comes back to the decision made by the individuals involved. The factors playing into those decisions varied; some of them were internal to the firm, some were not. However, only by directing attention to what people made what decisions with respect to the materials used and the objective to be obtained can we understand what happened. Unless it requires commentary, I will just go through the last five findings of the investigation.

The second item Colvin relates is

2. **Unexpected results from Inverse Null Corrector.** The inverse null corrector was used by Perkin-Elmer as a part of the reflective null assembly. It could be used as a check on the reflective null corrector's alignment and stability. It could show whether there was a gross flaw in the reflective null corrector as well as measure its stability. Since the inverse null emulated what the primary mirror surface should be when finished, it should have produced an interferometric pattern with straight lines, or in other words a null condition.

Instead, a pattern of wavy fringes was produced by the inverse null corrector at each testing cycle using the reflective null corrector. Although the inverse null corrector was producing results which did not agree with expectations, no nonconformance report was generated, and the designer was not consulted. While no qualitative analysis was performed, the inverse results were dismissed as being attributable to error in the null corrector's manufacture.

As a result of this condition, Perkin-Elmer managers decided to make a design change which revised the usefulness of the null corrector from its initial utility as a "double check" on the health and stability of the reflective null. Through testimony we ascertained that Perkins-Elmer managers decided to use the inverse null corrector solely as a stability check.

3. **Refractive Null and Reflective Null Tests Do Not Agree**. Within a week of the first inverse null results, Perkin-Elmer received other unexplained testing results. A refractive null corrector was used by Perkin-Elmer in Wilton, CT, for initial grinding of the mirror blank to get the mirror surface shape roughly to the desired finished shape. The mirror was then transported to Danbury, CT, where testing was first conducted using the reflective null corrector.

The test results from the first reflective null corrector test and last refractive null corrector test did not agree. Yet the expectation was the interferograms would be similar. In assessing the differences, Perkin-Elmer commented on the poor quality of the interferograms – caused by the rough surface of the mirror. They also asserted that some hand polishing of the outer edges of the mirror had occurred after the mirror left Wilton. No further quantitative analysis was performed.

4. **Recommended Gross Error Test Not Performed**. At the conclusion of the polishing phase, a Perkin-Elmer Vice President and General Manager

> formally requested an internal review of the primary mirror certification by senior scientists employed by Perkin-Elmer. This group, called the technical advisory board, held a review of the mirror fabrication and test results. On May 21, 1981, they recommended to the Perkin-Elmer Vice President that an independent test be performed on the mirror. The recommendation stated that "another test of the figure using an alternative method such as a Hartman test or a bean should be made. The purpose of the test would be to uncover some gross error such as an incorrect null corrector." No such test was ever performed.

If no such test was made, we can infer that the Perkins-Elmer Vice President decided not to make it. Why he decided that most probably has to do with the factors alluded to above, cost overruns and schedule delays.

> 5. **Vertical Radius Test Anomaly**. In May 1981, the most significant of the irregular events occurred when the refractive null corrector was used to measure the center of the curvature of the mirror . . . the refractive null interferogram showed wavy lines – clear evidence of an error in one of the measuring devices and possibly the mirror . . . the chance that two separate measuring devices, the inverse and refractive null correctors, would find matching errors and both be wrong and the reflective null right – is infinitesimal. . . . Yet Perkin-Elmer . . . failed to resolve the discrepancies in quantitative ways . . . had Perkins-Elmer attempted to determine the source of the error, with the analytical and measurement tools in place at the time, they could have determined that the flaw was in the reflective null corrector in one or two days. Perkin-Elmer personnel assumed erroneously that there were large "as built" errors in both the refractive and inverse null correctors. They ignored the results of the intended sanity checks on the reflective null corrector . . . the results of the last . . . test concerned Perkin-Elmer managers, but they did not disclose the results or their concerns outside of the optics fabrication group . . . to our best determination, Perkins-Elmer did not share the discrepant results of the vertex test with NASA. The NASA Plant Representative . . . was provided a copy of the center portion of the vertex radius test interferogram. This cropped version did not disclose the curved fringes which would have indicated a problem with the test results.

Here we are at a crucial point. Someone deliberately cut out the center of the interferogram in order to not reveal evidence of a problem. The decision to hide relevant evidence is at the heart of the resulting problem with HST since, as we shall see, other members of the Perkins-Elmer group were left in the dark. Colvin's report continues:

> Perkin-Elmer quality assurance personnel told us that they were present at daily meetings on the HST and they were never made aware of aberrant test results, nor were there any such discussions at the meetings.

If this last point is true, then it suggests that communication within the company was flawed. It is not clear if there was competition or merely an effort to not look bad. It begins to sound like upper management charged with overseeing manufacture of the mirror dug itself into a hole which further decisions simply deepened. I am not sure if this supports or undermines Bucciarelli's view.

Finally, there was rapid close-out of the Perkins-Elmer optical fabrication team. The program manager had eleven items he wanted to check out, but was not given approval to do so. And in conclusion, "[T]he Head of Manufacturing Optical Analysis stated that he and his manager would have couched the vertex radius anomaly in terms of a need to 'recertify the reflective null corrector'." This is nothing more than an admission of intended obfuscation, which tends to support my conclusion above.

5. CONCLUSIONS

What are we to make of this incredible list of failures to report bad test results, of failure to adhere to design specifications, and failure to adhere to protocols? It might be attributable to the internal ethos at Perkins-Elmer. But it seems there were other factors at work as well. Robert Smith (1989) details the myriad of political pressures that were also at work. Economic considerations were paramount. Virtually every decision in the early days through the mid-eighties was made to lower costs. Everyone wanted a space telescope, no one wanted to pay for the real McCoy. The decision to launch the HST on the shuttle forced further design changes that aggravated the complexity of the project. At every turn, as we uncover layer after layer, we find that it is the decisions that were made, motivated by varying considerations from political clout to economics to engineering design, that give us the clues to what actually happened. This is sketchy – the case itself is complicated because the technical nature of the design needs to be part of our conversation, as does the political context and the economic situation. For example, budgetary considerations forced a cut-back in the number of NASA inspectors assigned to the project. At every turn we find crucial decisions based on non-uniform considerations that point the way. Finally, we find that failure to assess decisions in light of the larger scope of the project are what ultimately led to failure. Why bad tests were not reported and discussed up and down the line is not clear. Why designers were not consulted when production problems emerged is not clear. What is clear is that the resulting engineering design failure was a failure to utilize the feedback function of my decision-making model.

It is important here to pause to consider a possible objection. It might be argued that the launching of the defective Hubble was not an engineering failure so much as one of management. Clearly the final decisions not to test and to ignore the results of tests that were already on record was a management decision. But the discussion here is not about engineering versus management. It concerns the design process. This was a government contract and it had a number of specific protocols built into it. These included consulting with designers when problems occurred, using the null refractor results as intended, etc. What happened was clearly a breakdown in the design process, a process which included both line engineers and management. To claim that this was all a management problem ignores the fact that it all began with some technicians failing to follow protocol. My concern is how the design process actually works. What we have here is an example of a process that is set out in writing and then ignored by virtually all parties.

In looking at Bucciarelli's and Vincenti's models, with this case in front of us, several things can be noted. First, while Vincenti's structured account of the design process puts us on the road to systematic analysis, by leaving out people, he omits the mechanism for moving back and forth between his various stages, i.e., the decisions. People are presupposed and that is not enough. Without specifically including people in the structure, we cannot focus on their decisions, and it is clear that it is the decisions that form the loci for understanding.

As before, Bucciarelli's move to the social takes us in the right direction, but I fear his work by virtue of objectifying the object world in a way delimits human agency. But as it should be clear from the HST case, human decisions are at the heart of what happens to the object.

Finally, it is also clear that we cannot argue that these were just bad engineers who were responsible for the problems with the HST; that is, unless you build honesty, apple pie and the American Way into your definition of a good engineer. These folk knew enough to engage in systematic behavior designed to hide their mistake and they succeeded, until someone turned on the telescope.

What can we glean from this brief look at this case? If there is a moral it is this: the main focus of efforts to understand technological projects needs to be the people doing the work. It is their decisions that hold the promise of understanding how what happened happened. True, those decisions will be contextualized by a variety of factors, but few of them can be determined ahead of time except in the most general terms – people, materials, institutions, etc. – but that tells us little. The devil is in the details; grand schemes seem increasingly less appealing. This also applies to MT. All we really need to work on is a better understanding of why people do what they do. If MT helps at all, it is in breaking down the process of human activity into steps for analytic purposes. Perhaps this is what the empirical turn in the philosophy of technology is all about.

REFERENCES

Bachelard, G. (1934). *Le Novel esprit scientifique*. Paris: Alcan.

Bucciarelli, L. (1994). *Designing Engineers*. Cambridge, MA: MIT Press.

Chaisson, E. J. (1994). *The Hubble Wars*. New York: Harper-Collins.

Colvin, W. (1993). Testimony in the *U. S. Congressional Record*, November 16, 1993, 36–39.

Hanson, N. R. (1968). *Patterns of Discovery*. New York: Cambridge University Press.

Pitt, J. C. (1987). The Autonomy of Technology. In: P. T. Durbin (Ed.). *Technology and Responsibility, Philosophy and Technology, 3* (pp. 99–114). Dordrecht: Kluwer.

Pitt, J. C. (1995). Discovery, Telescopes and Progress. In: J. C. Pitt (Ed.). New Directions in the Philosophy of Technology, *Philosophy and Technology, 11* (pp. 1–16). Dordrecht: Kluwer.

Pitt, J. C. (2000). *Thinking About Technology: Foundations of the Philosophy of Technology*. New York: Seven Bridges Press.

Rudner, R. (1953). The Scientist qua Scientist Makes Value Judgments. *Philosophy of Science, 20*, 1–6.

Schubert, G. (1965). *Judicial Policy-Making*. Glenview: Scott, Foreman.

Smith, R. (1989). *The Space Telescope*. New York: Cambridge University Press.

Vincenti, W. (1989). *What Engineers Know and How they Know It*. Baltimore: Johns Hopkins University Press.

Winner, L. (1986). *The Whale and the Reactor: A Search for Limits in an Age of High Technology*. Chicago: University of Chicago Press.

PART IV

ETHICS AND THE EMPIRICAL TURN

THE ETHICS OF ENGINEERS: FROM OCCUPATIONAL ROLE RESPONSIBILITY TO PUBLIC CO-RESPONSIBILITY

Carl Mitcham and René von Schomberg

Empirical turns in the philosophy of technology take place onto more than one roadway. From the broad thoroughfare of philosophy any turn onto the more narrow highway of ethics (in contrast to the avenue of epistemology or the boulevard of metaphysics) is sometimes interpreted as a kind of empirical detour. Yet surely ethics must be part of any empirical turn in the philosophy of technology broadly construed.

From ethics itself one may also take empirical turns – not only from metaethics and normative ethics to descriptive ethics, but also from ethical theory to ethical codes of conduct, for example. Indeed, in the professions associated with technology – meaning primarily and foremost the engineering professions – discussions of ethics are almost always related, quite empirically and experientially, to questions of the ethical responsibilities associated with the role of being an engineer.

The emergence of philosophically associated engineering ethics thus constitutes in itself what may well be denominated as an empirical turn in the philosophy of technology.[1] What follows is an attempt to give some descriptive content to this turn, and to raise – but not without empirical references – some appropriately philosophical questions about it.

The Empirical Turn in the Philosophy of Technology, Volume 20, pages 167–189.
2000 by Elsevier Science Ltd.
ISBN: 0–7623-0755–2

1. ROLE RESPONSIBILITY: HISTORICO-PHILOSOPHICAL BACKGROUND

Engineering ethics is a robust example of role responsibility ethics. It is not so much based on ethical theory or ethical principle as it is on a highly descriptive and empirical examination of roles, context, and their implications for conduct.

According to a standard social science definition, the term "role" refers to "the behavior expected of the occupant of a given position or status" (Sarbin, 1968, p. 456).[2] "All the world's a stage/and all men and women merely players," wrote Shakespeare (*As You Like It* II, 7). Elaborating on this theatrical metaphor,

> an actor assigned to the position (or part) of Hamlet is expected to enact the *role* of Hamlet, the role being characterized by certain actions and qualities. A person who is assigned to the position of clergyman (or who elects to be placed in such a position in the social structure) is similarly expected to enact the *role* of clergyman . . . (Sarbin, 1968, p. 546).

The concept of role as a social or occupational position thus combines descriptive and normative elements. Given, in one way or another, that an individual takes on some socially structured role, especially of a professional character, such an individual becomes subject to certain representations and responsibilities. The responsibilities involved will inevitably have strongly ethical significance.

The idea of role responsibility in ethics can be traced back at least to Plato's *Republic*. There Socrates, having gone down to the Piraeus in curiosity about a new religious festival, is drawn into an evening of conversation about justice. This conversation ultimately leads up to an expanded description of a hierarchical state in which are distinguished three types of members: rulers, guardians, and artisans. Each class has its distinctive role to play, and justice is defined as the members of each class meeting their particular role responsibilities. Aristotle's remarks in the *Politics* about the naturalness of certain roles may be read as the pursuit of empirical warrant for Socrates' nocturnal ascent. Epictetus, a Stoic follower of Socrates, further endorses acceptance of role responsibility as something akin to a spiritual exercise (Epictetus, *Encheiridion* xvii).

Because of the political character of such roles, their boundary conditions are often defined and enforced by law. Although not explicitly detailed in either the *Republic* or *Politics*, certainly in many cities there have been written codes and statutes governing not only participation in the military but also, for instance, prohibiting artisans from infringing on one another's areas of competence. The well-known *Code of Hammurabi* (c. 2000 BCE)) even goes so far as to establish detailed legal boundary conditions on the practice of many arts, including those of the builder of houses (see, e.g., *Code of Hammurabi* 228–233).

In the classical tradition, Cicero's *De officiis* provides the most well-developed and personalized account of role responsibility.[3] (The Latin *officium,* etymological root of the English "office," although commonly translated as "duty" or "responsibility," itself has a strongly role-based connotation.) According to Cicero, each person is endowed by *natura* or nature with two *personae* or roles (*De officiis* I, xxx, 107): that which is common to all and distinguishes humans from animals, namely, the role of being rational; and that which is particular to each individual, namely, the role of being in a particular body with its specific temperament.

These natural roles are supplemented by two others (*De officiis* I, xxxii, 115): one historically contingent, another voluntarily chosen. A person is historically born into the role of Roman or non-Roman, rich or poor; one chooses to take on the role of philosopher, jurist, or orator – and to excel in one virtue or another. According to Cicero, *Haec igitur omnia, cum quaerimus quid deceat, compleciti animo et cogitatione debemus* . . . All [four of] these, when we inquire into what is decent [or proper], we ought to engage in spirit and thought . . ." (*De officiis* I, xxxii, 117).

In short, in order to achieve what Cicero refers to as *decorum* in one's responsibilities (*De officiis* I, xxxi, 110), one should take into account what might be called the four-fold root of role responsibility. In one sense, the role of exercising rational liberty in choosing roles is primary, although most difficult; in another, each human being must be careful to respect and live in harmony with what is given by nature and by history, both to the species and to the individual. For Cicero responsibility is not simply accepting a social role, but involves as well a grounding of that social role in nature and history – and the deepening that comes from reflective attention to a matrix of role possibilities.

In the Christian tradition the notion of role responsibility continues to be affirmed, but with a degree of detachment that intensifies the voluntary component already identified by Cicero. According to St. Paul, for instance, in Christ "there is neither Jew nor Greek, slave nor free, male nor female" (Galatians 3: 28). At the same time, Christians should "be subject for the Lord's sake to every human institution" (1 Peter 3: 13), and live out their appointed roles.

Within Christianity, the traditional social role responsibilities, defined by exclusively human institutions and their laws, begin to be interpreted in a radically contingent manner. There cease to be, as with Cicero (and before him, Plato and Aristotle), any unalterably natural roles. In place of the primary distinction between natural and conventional roles there arises one between unconsciously inherited roles (mores or customs) and consciously willed ones. For Christians, what is crucial is the distinction between those who live by the law and those who live by faith, that is, between those who have not and those who have consciously taken upon themselves and professed the name of Christ, and as a result live "in the world but not of it."[4]

Within Christian theology there nevertheless emerge new role responsibility distinctions based on what are termed different "states of life": those who have taken holy orders (members of the priesthood) and the laity. For the laity in turn, there are another two possibilities: the single state and the married state. Intermediate between the laity and the priesthood is the religious state, that is, those who have taken vows to follow not simply the Christian commandments (love of God and of neighbor as oneself; see, e.g., Matthew 22: 34–40) but also the counsels (of poverty, of celibacy, of obedience; Matthew 19: 10–12 and 21). Within the carefully articulated social order of the religious community (see, e.g., the *Regula* of St. Benedict), role responsibility remains the background of an increasingly refined ethics of principle and conscience.

Although modern Western philosophy explicitly rejects any Christian theological ties, the distinctly modern tradition of ethics largely continues to presume the radically contingent character of all social roles and, as a result, to focus attention on movements from custom and law to ethics, with a consequent emphasis on responsibility according to principle. It is differences over principles, for instance, that divide consequentialism (John Stuart Mill and associates) and deontologism (Immanuel Kant and associates). It is divergent principled responsibilities that pluralists (such as W. D. Ross) try to integrate. Discussion of role responsibility has, especially in the Anglo-American world, been turned over to the more empirical sciences of sociology and psychology, although some American philosophers in the pragmatist school (such as George Herbert Mead) reintroduced the notion into social philosophy.[5]

In modern philosophy, however, the most sustained attempt to defend and develop role-responsibility ethics is associated with G. W. F. Hegel. In explicit opposition to Kant's high principlism, Hegel develops an extended analysis of *Sittlichkeit*, commonly translated as "ethical life," as the phenomenological foundation for any particular invocation of rational moral principles. Hegel's *Grundlinien der Philosophie des Rechts* (first published 1821) argues for what has become known as an organic theory of the state, in which broad role differentiations between family, civil society, and state establish regions of responsibility and the possibilities for its further articulation. Like Plato's Socrates of the *Republic*, Hegel views one's role in a social order as providing the framework for ethical behavior; like Cicero, he sees that individual roles may be multiple and subject to development – both personally and historically.

In English the most sustained version of the associated role responsibility ethics is found in the British Hegelian F. H. Bradley's *Ethical Studies*, especially chapter five, "My Station and Its Duties." Between the aggregate subjective hedonism of utilitarianism and the abstract moral absolutism of deontology, Bradley argues that real morality is found in "our station and its duties, our function as an organ in the social organism" (Bradley, 1911, p. 148).

For Bradley the ultimate end of human action is self-realization. But what is to be realized in and through the self is neither the brute individual, in which there is only subjectivity with no higher aim than self-satisfaction, nor the abstract universal, which is an aim so high as to be only thinkable. Concrete human beings are born into complex social wholes in which they learn to exercise freedom by playing their parts. The concrete level of self-realization is thus neither the individual pursuit of self interest nor the abstract attempt to conform one's actions to a purely rational ideal, but the effort to live up to one's social role – which at once raises individuals above themselves and brings the abstract universal down into particulars.

Social roles both represent the natural complexity of reality and provide ideals that realize a self. In Bradley's words,

> I am myself by sharing with others, by including in my essence relations to them, the relations of the social state. If I wish to realize my true being, I must therefore realize something beyond my being as a mere this or that; for my true being has in it a life which is not the life of any mere particular, and so must be called an universal life. What is it then that I am to realize? We have said it in "my station and its duties." To know what a man is . . . you must not take him in isolation . . . . What he has to do depends on what his place is, what his function is, and that comes from his station in the organism (Bradley, 1911, p. 157).

In the 20th century the philosophical credibility of organic role responsibility ethics has been undermined by its close association with traditional religious conservatism (Hindu caste system and Christian racism) and political totalitarianism (fascism and communism). Until the quite recent but only marginally acknowledged ascendancy of role responsibility ethics in the form of professional ethics,[6] the only contemporary moral theorist of stature to have anything positive to say about role responsibility has been Alasdair MacIntyre, a philosopher with strong if querulous ties to the social sciences.

In his history of ethics, for instance, MacIntyre notes that "when we detach a man from his roles, but still leave him with the concept of 'duty,' the concept is necessarily transformed" (MacIntyre, 1966, p. 93). Commenting more specifically on this transformation as it took place in modern English society:

> Originally we have a society of well-defined occupational roles and functions, hierarchically arranged, and to this arrangement there corresponds a belief in different stations in life to which God is pleased to call men. When the occupational roles become more important, the notion of a calling by God, but not to any one particular "station," remains . . . In such a situation the content of duty will be blurred. [There is] a move from the well-defined simplicities of the morality of role fulfillment, where we judge a man *as* farmer, *as* king, *as* father, to the point at which evaluation has become detached, both in the vocabulary and in practice, from roles, and we ask not what it is to be good at or for this or that role or skill, but what it is to be "a good man"; not what it is to do one's duty as clergyman or landowner, but as "a man." The notion of norms for man emerges as the natural sequel to this process, and opens new possibilities and new dangers (MacIntyre, 1966, pp. 94–95).

Indeed, one reading of MacIntyre's subsequent argument for a revived attention to the importance of the virtues in ethics (MacIntyre, 1984) is to see it as an attempt to deal with the new dangers.[7] Virtues may function much like roles, but are nevertheless not socially defined.

2. THE RISE OF PROFESSIONAL ENGINEERING AND ITS ROLE RESPONSIBILITIES

Although the philosophical defense of role responsibility has fallen into disfavor, what philosophers sometimes describe as the regionalization of ethical principles in many of the areas of applied ethics – biomedical ethics, computer ethics, etc. – actually constitutes, as it were, a kind of empirical revival of responsibility in relation to roles. Indeed, this surreptitious resuscitation of role responsibility has taken place mostly from the ground up, with little assistance from professional philosophy. Consider, for instance, the case of engineering ethics and especially the formulation of engineering ethics codes.

The role-defining codes of conduct for professional engineers have been formulated in conjunction with the development of engineering as a technical professional occupation, which itself began at roughly the same time as the classical emergence of the Industrial Revolution. (This was also precisely the period during which, on MacIntyre's account, social roles began to be replaced by occupational ones). This technical professional development may be summarized in terms of four historical phases:

Phase one: The creation of specialized schools of technical engineering education. Such schools are represented by the École Nationale des Ponts et Chaussées (founded 1747), the École Polytechnique (1795), and the U.S. Military Academy at West Point (1804). In this phase, because all engineering was state or military engineering, the unwritten code of conduct for engineers was determined by their roles as civil servants or soldiers. The basic role responsibility of soldiers is to obey orders; together with technical engineering competency this role responsibility was inherited by the engineering corps.

Phase two: The creation of professional engineering societies independent of the military. Examples include the British Institution of Civil Engineers (granted a royal charter in 1828) and Institution of Mechanical Engineers (1847), the American Society of Civil Engineers (ASCE, in 1852) and American Society of Mechanical Engineers (ASME, in 1880).[8]

The formation of professional societies has continued to follow the emergence of technical engineering disciplines. Examples include

– the American Institute of Electrical Engineers (AIEE), in 1884;

- the Institute of Radio Engineers (IRE), in 1912;
- the merger of the AIEE and the IRE in 1963 to form the Institute of Electrical and Electronic Engineers (IEEE), the largest professional engineering society in the world; and
- the Association for Computing Machinery (ACM), in 1947.

The earliest such professional associations functioned primarily to promote technical competence and solidarity among non-military engineers. Having stepped out of a well defined military role into a less structured civilian milieu, professional engineers experienced mutually reinforcing needs to affirm occupational role distinctiveness and to develop technical competence. Although technical competence is the *sine qua non* of engineering as a profession, in order to exchange technical information engineers also found it useful to create social organizations, which in their turn both required and reinforced social identity.

Phase three: Formulation of the first explicit codes of conduct. The explicit adoption of codes of conduct by professional engineering societies did not take place until early in the 20th century. When it did, it focused on what may be termed internalist issues of occupational role responsibility: loyalty to the profession and to employers, the maintenance of technical knowledge and skill, avoidance of conflict of interest, etc. Such obligations are, for instance, the primary obligations of the 1912 code of the AIEE and the 1914 codes of both the ASCE and ASME.[9]

The very need for explicit codes no doubt derived in part from the increasing size of the engineering profession. In small groups personal apprenticeship suffices to pass on the ethos of a profession. Moreover, the emphasis on loyalty, although stated as though to a client, actually functioned – since most engineers were salaried employees rather than independent consultants – as company loyalty. Company loyalty has been interpreted as simply a civilian version of the obedience to authority formerly operative among military engineers (Mitcham, 1987).[10]

During this same period, however, some engineers attempted to go beyond notions of company loyalty and to identify in their technical occupation itself a defining ethical ideal. Just as the role of physician is governed by the medical ideal of health and that of the lawyer by legal justice, so it was argued engineering pursued an ideal at once technical and moral, descriptive and prescriptive: efficiency.[11] During the first third of the 20th century such an interpretation of efficiency even gave rise to attempts to develop technocracy as a positive social-political ideal (see, e.g., Akin, 1977). Indeed, the concept of efficiency continues to exercise a two-sided, technical-moral influence not only in engineering but also in economics and in politics.

Phase four: Ethics code revision. In response both to further changes in the profession itself and to transformations of the societal context within which engineers work, there began in the middle of the 20th century to take place a revision and reformulation of ethics codes. Internally, as a result especially of World War II, science became what Derek de Solla Price has simply (and aptly) termed "big science" (Price, 1963); engineering likewise became "big engineering". Externally, the late 1950s and early 1960s witnessed the emergence of widespread social criticism of engineering as a cause of environmental pollution and a challenge to the quality of material culture.

As a result of both these internal and external changes, technical professional codes of ethics, especially in the United States, while continuing to stress internal occupational role responsibility, began to pay at least nominal attention to externalist issues. For instance, in 1947 the Engineer's Council for Professional Development (ECPD, founded 1932) drew up the first trans-disciplinary engineering ethics code, and in the process committed engineers "to interest [themselves] in public welfare." Further revisions in 1963 and 1974 strengthened this commitment to the point where the first of four "fundamental principles" required engineers to use "their knowledge and skill for the enhancement of human welfare," and the first of seven "fundamental canons" stated that "Engineers shall hold paramount the safety, health and welfare of the public . . ."

As one other example, when the code of ethics of the newly created IEEE was finally articulated in 1974 it included in the fourth of four articles a responsibility to "protect the safety, health, and welfare of the public" and to "speak out against abuses in these areas affecting the public interest." In a 1987 revision of the code this article was effectively if only minimally elevated by the addition of a fifth article after it. But then just three years later in 1990, in a more fundamental revision, the code was simplified to ten much shorter principles, the very first of which is "to accept responsibility in making engineering decisions consistent with the safety, health and welfare of the public."

Some version of this public responsibility plank is now included in all major U.S. engineering ethics codes. This empirical fact nevertheless deserves to be qualified by two critical observations. The first observation is that although public responsibility has been included in engineering ethics codes, this has not necessarily made it a fully active ingredient of such codes. The plank is deeply problematic on at least three interrelated counts:

(a) It is difficult to promote and virtually impossible to enforce. Indeed, mechanical engineer and philosopher Taft H. Broome, Jr., has gone so far as to argue that difficulty of enforcement is grounded in impossibility of practice. The problem is that engineering inherently brings about risks.

> [E]ngineers *cannot assure* the health and welfare of the public because risk-free engineering often cannot be achieved. Thus a problem is posed for the incorporation of a public paramountcy rule into codes of ethics since any rule stipulating what one *should* do has no logical status if one *cannot* do it (Broome, 1989, p. 4).[12]

(b) It presumes individual actions producing well-defined and unique results. In fact, most engineers now work in teams, so that actions are collective even at the technical level, and their actions often have multiple unintended consequences. Indeed, so much are multiple unintended consequences the standard result that the most widely used engineering ethics textbook (Martin & Schinzinger, 1983) describes engineering as, in effect, social experimentation.[13] The power and complexity of contemporary technoscientific action, which because of its large-scale consequences calls for public accountability, nevertheless tends in its complexity to undermine the very possibility of such accountability.

(c) What actually constitutes public safety, health, and welfare is almost always contestable and subject to extensive public debate. The engineering of nuclear weapons is considered by some to be a means for protecting (national) public safety, health, and welfare; by others nuclear weapons are argued to be a threat to (international) public safety, health, and welfare. Take almost any major engineering project and a similar spectrum, if not divide, emerges.

The second critical observation regarding professional engineering attempts to promote public responsibility is that professional technical societies are not all there is to ethics in science and engineering. Professional ethics developments in the professional technical societies have been supplemented in major ways by professional activist associations such as

- the Federation of American Scientists (FAS, founded 1948) and International Pugwash (founded 1957), both of which arose to lobby for the limitation and international control of nuclear weapons; and
- Computer Professionals for Social Responsibility (CPSR, founded 1985), which arose in response to the perceived dangers of President Ronald Regean's Strategic Defense Initiative.

These and related activist associations of professional scientists – such as the Committee on Scientific Freedom and Responsibility (CSFR, established 1975) of the American Association for the Advancement of Science and the Union of Concerned Scientists (UCS, founded 1969) – clearly understand professional role responsibility ethics to include more than internalist occupational role responsibility issues, and to see public responsibility as more than nominal. For members of the FAS and CPSR, responsibility includes public, not simply technical, action for social responsibility. Such activity requires engineers to become involved in science and engineering policy making processes and even calls for public political activism. They point beyond occupational role responsibility as it has traditionally been conceived and toward what we will term collective co-responsibility.

3. CONTEMPORARY TECHNICAL ROLE RESPONSIBILITY AND ITS DISCONTENTS

What inferences may be drawn from this double sided historical background: the modern philosophical subordination of role responsibility to principle responsibility, and the counterpoint re-emergence and development of engineering role responsibility – especially of a problematic engineering occupational (as opposed to social) role responsibility for public safety, health, and welfare?

The historical shift from social role to ethical principle to the re-emergence of role in the form of occupational role is part of a much larger historical replacement of an embedded or integrated social role-based lifeworld with a world of well differentiated systems. Our point of reference here is the work of Jürgen Habermas. Drawing on the classic founders of modern social theory such as Karl Marx, Emile Durkheim, and Max Weber, and refining the thought of Talcott Parsons, Habermas provides the most extended analysis of the devolution of the traditional lifeworld into a systems world – or, in alternative terminology, the colonization of the lifeworld by instrumental rationality. In Habermas words, "the uncoupling of system and lifeworld" takes place

> in such a way that the lifeworld, which is at first coextensive with a scarcely differentiated social system, gets cut down more and more to one subsystem among others. In the process, system mechanisms get further and further detached from the social structures through which social integration takes place. [M]odern societies attain a level of system differentiation at which increasingly autonomous organizations are connected with one another via delinguistified media of communication: these systemic mechanisms – for example, money – steer a social intercourse that has been largely disconnected from norms and values, above all in those subsystems of purposive rational economic and administrative action that . . . have become independent of their moral-political foundations (Habermas, 1987, p. 154).

What Marx and Durkheim had initially analyzed as a process of increasing division of labor and specialization emerges in society at large in three forms. One is the differentiation of societal subsystems of business enterprise (economy), public administration (politics), and law (as an integrative subsystem). Two is the emergence of science, art, ethics, and related forms of knowledge as autonomous realms of culture, followed by their disciplinary and subdisciplinary occupational professionalizations. Third is the transformation of the general social roles of wage laborer and citizen into those of consumer and voter or welfare state client.[14]

Within such a sociological framework, empirically speaking it is not principle-based ethics but occupational role responsibility ethics on which the organization of contemporary techno-social life fundamentally depends. This point is confirmed by the functional prominence of role responsibility ethics in techno-scientific fields, as illustrated by engineering ethics. At the same time, occupational role responsibility is being stretched to the breaking point by its own intensification, as may be summarized in terms of four late 20th and early 21st century trends and three often rehearsed responses.

First, there has occurred, as a consequence of the professionalization of multiple tasks previously carried out in non-technical or private spheres, an enormous differentiation of new roles. Engineering itself provides modest illustration, as it has broadened its functional specializations from research, development, design, and construction to include production, operation, management, and even sales engineers; and its content specialization to include biomechanical engineering, biomedical engineering, biochemical engineering, nanoengineering, and more. Stepping outside technical fields, the unfortunate *reductio ad absurdum* in this trend is the role professionalization of virtually every work-related activity: janitors become maintenance professionals, friends are replaced by professional psychological counsels, one hires professional personal trainers to help one get the right exercise, etc. Although this development is primarily manifest as the quantitative proliferation of roles, it inevitably has qualitative implications (see, e.g., Illich et al., 1977).

Second, and in parallel, the area for which an individual may be held responsible has been narrowed, as may be illustrated with an example from the sciences. In the 1700s there were natural philosophers who pursued natural science. In the 1800s William Whewell coined the term "scientist," and initially there were simply scientists as such (separate from philosophers). This was followed by a period in which it was possible be a physicist, chemist, or biologist. Today, however, not even the term microbiologist is sufficiently descriptive of a scientific role. As a result some individual scientists may only be proficient in research they conduct on one specific microorganism, perhaps solely in relation to a restricted number of biochemical processes in that microorganism. Individual scientists increasingly "know more and more about less and less," and thus can hardly foresee the consequences of their discoveries for related fields, let alone the possible applications that could result from interactions with other fields. Such an excessive differentiation of roles implies both a formal and a substantial delimitation in individual role responsibility.

Third, the number of roles that any one individual may possibly fill has dramatically increased. Synchronically, one person may well be a structural engineer (that is, a kind of civil engineer) doing research on earthquake remediation, a grant or contract administrator, a professor of engineering, a student advisor or mentor, an academic administrator (as department head or dean), an author – not to mention a spouse, parent, church member, citizen, consumer, and more. Diachronically, the same person may alter all of these roles and/or complement them with literally hundreds of others. Indeed, the interchangeability of individuals and roles has expanded along with individual mobility, both temporally and geographically. This means, practically, that responsibility is more identified with a role than with a person, thereby complicating the responsible organization of professional tasks while significantly diminishing technical professional ethical commitments – not to mention loyalty.

Fourth, contemporary society is not only characterized by the differentiation of roles but also by the intensified institutionalization of the social-institutional spheres in which the role differentiation takes place. Science, engineering, economics, education, politics, art, religion, and more have all become so institutionally distinct that they largely determine the conditions for their own functioning. Regulation, insofar as it occurs, must increasingly take place internally within each sphere. Scientists regulate science, engineers engineering, and so on.

As a result of this four-dimensional transformation of role differentiation space, technical roles may be said to have become increasingly less robust at the same time that opportunities for role conflict have only intensified, proliferated, and specialized, with individuals more freely floating between roles, although large role aggregates are more rigidly separated from each other than ever before. The result is a multifaceted undermining of that very role responsibility that is the basis of social order – and for which it is dubious that principle responsibility alone is able to compensate.

Although roles are central to the functioning of technoscientific society, technical responsibility is progressively weakened in the moral sense. During the last half of the 20th century in contemporary technological societies, professional roles gained such prominence that, together with their associated expectations and codes of conduct, they constitute one of the major foundations of contemporary ethical problems and dilemmas. Especially the role responsibility of executing assigned tasks from superiors has, outside of professional philosophy, become an important ethical issue.

As was most dramatically demonstrated in the 1962 trial of Adolf Eichmann, strict adherence to occupational role responsibility easily leads to an almost banal immorality (Arendt, 1963). During his trial, Eichmann defended himself by appealing to his role as chief administrator of the mass execution of Jews during Word War II, pointing out that his responsibilities were limited to administrative tasks in a hierarchy in which he had to fulfill the orders and follow the instructions given to him by superiors. Although the Eichmann case is exceptionaly horrifying, the kind of appeal he made is not so exceptional at all. Repeatedly individuals in technoscientific and contemporary management positions find themselves resorting to precisely this line of reasoning to justify their behaviors. The widely studied Challenger disaster of 1986, for example, may readily be interpreted as illustrating this phenomenon (see Vaughan, 1996).

This infamous example and its not-so-infamous parallels have not, however, led to any wholesale rejection of individual occupational role responsibility ethics. Instead, in the first instance it is often used to argue that individuals must simply acknowledge more than one administrative or technical role. Discussion has therefore focused more on the ethical dilemmas and conflicts that arise when two or more occupational roles conflict.[15] This has varied from an emphasis on conflicts between the roles of being the member of a family and a professional to issues of the extent to which a technical professional may in certain situations have a responsibility to become a whistleblower. Rather than leading to examination of the adequacy of occupational role responsibility itself or of the contemporary role differentiation space, the dilemmas of role responsibility have themselves simply become the focus of discussion. To resolve these dilemmas within an occupational role responsibility framework has been the primary intellectual concern, rather than to challenge the ethics of occupational role responsibility itself.

A second attempt to deal with what may be termed the Eichmann problem has been not just to argue for the acknowledgement of more than one technical role but to actually increase technical role responsibility. The historical development of engineering ethics in the United States well illustrates this response. Engineers are not simply responsible for efficiency, nor are physicians single mindedly devoted to health and lawyers to legal justice, but all are as technical professionals in some sense charged with upholding the public good. The problem, with regard to engineers, is that it seems virtually impossible to provide technical-professional mechanisms which would enable individuals to meet this responsibility for public safety, health, and welfare. Those who take this issue seriously tend to spill over into activist organizations such the Federation of American Scientists and Computer Professionals for Social Responsibility – which, as previously suggested, point beyond occupational role responsibility.

Still a third approach to address the role responsibility problem has involved attempts to develop an "ethics of technology" (see especially Jonas, 1984) or "ethics of science,"[16] as well as a variety of studies that typically build on the phrase "social aspect of" in their titles – e.g., the social aspects of engineering, the social aspects of computing, etc.[17] Such fields of interdisciplinary scholarly activity are, however, more concerned with exploring and cataloging the phenomena themselves than with the underlying social orders or the development of normative responses to the occupational role responsibility problem itself.

Interdisciplinary studies of the ethics of science and technology nevertheless regularly highlight the extent to which people increasingly feel inadequate to deal with the complex moral dilemmas in which role responsibility places them. The more common phenomenon, in the face of Eichmann-like situations, is not Eichmann-like self justification, but what Austrian philosopher Günter Anders (Eatherly & Anders, 1961) might associate with the doubt and guilt manifested by "Hiroshima bomber pilot" Claude Eatherly.[18] But was Eatherly really responsible? What about J. Robert Oppenheimer, the leader of the scientists and engineers who designed the bomb? Or what about President Harry Truman, who ordered the bomb dropped? Or President Franklin Roosevelt, who established the Manhattan Project? Or even Enrico Fermi and Albert Einstein, who wrote the 1939 letter to Roosevelt that called attention to the possibility of an atomic bomb?

The very complexity of the atomic bomb project calls into question any attempt to accept personal responsibility for the results. Yet certainly Oppenheimer and many other atomic scientists experienced some guilt, and their concerns led to the kinds of public activism illustrated by the founding of the Federation of Atomic (later American) Scientists and the creation of the *Bulletin of Atomic Scientists*. Anders' paradoxical critique and idealist call for expanding human powers of imagination and responsibility (see also Anders, 1980) is but the more philosophical manifestation of that intensification and multiplication of moral dilemmas which has led many people to feel that various issues are at once their responsibility and/or beyond their role competencies. The familiar not-in-my-backyard (NIMBY) syndrome in response to industrial construction or waste disposal and personal refusals to limit the consumption of high pollution consumer goods such as automobiles are but two sides of the same coin.

What thus emerges from our description of this four-dimensional transformation of the technical role responsibility space and the three attempts to respond to such a transformation is the picture of a society in which there is an imbalance between the individual's responsibility in temporary occupational roles and the collective responsibility which is represented by the simultaneous fulfillment of great number of roles over the long-term. This is illustrated by the fact that in increasing numbers of instances it is impossible, even in a hierarchically structured technical professional system to assign to any one person responsibility for solving some particular problem. Who or what role is responsible for nuclear weapons proliferation? For stratospheric ozone depletion? For global climate change? Indeed, who or what role is responsible for even such mundane problems as traffic congestion? For the malfunctioning of my computer? For the presence of unlabelled genetically modified foods in grocery stores? The chance that any one individual can be identified as responsible for the consequences of our collective actions within and between the myriad systems and subsystems of the technoscientific world has become infinitesimally small. Instead, in most instances it is increasingly the case that some form of co-responsibility for a collective organization and action leading to consequences (both intended and unintended) is operative. At the same time, such collective co-responsibility is difficult to grasp and elusive; it often seems as difficult to pin down some individual, organization, or even single system that might be held accountable for numerous scientific and engineering developments.

4. FROM INDIVIDUAL ROLE RESPONSIBILITY TO COLLECTIVE CO-RESPONSIBILITY

We have described, in an admittedly summary manner but with strong empirical references, a society in which it is difficult for anyone to be held responsible for the consequences of many technoscientific actions. We rely on a theory of occupational role responsibility that is no longer in harmony with existing social reality, in response to which we commonly propose an alternative and expanded notion of role responsibility. The fact is that the consequences of a wide variety of collective actions cannot be reconstructed from the intentions of responsible individuals, and role responsibility ethics can bear only on the consequences of individually and intentionally planned actions.

Individuals assume responsibility for the consequences of their actions if and only if they can intentionally direct those actions and reasonably assess the consequences, both intended and unintended. (Unintended consequences may on some occasions be effectively covered by insurance, as with automobile insurance.) But the consequences of scientific discovery and engineering design often escape all common or natural means of assessment.

Science and engineering exist, in the first instance, within the scientific and technological systems and, subsequently, by means of a complicated transformation and use, are transplanted into the system-specific logics of the economy, politics, and law. None of these system logics are traceable to the intentions of individuals, nor are the possible unintended consequences always assessable. Scientists who have knowledge that leads to applications which are then criticized by many in society, may rightly point out that they anticipated other applications. Engineers who design products, processes, or systems that wind up actually being used in a variety of ways (guns that kill people as well as protect them, for example) make the same argument. Scientists and engineers may even claim that the possible applications and/or uses are not part of their occupational role responsibilities as scientists or engineers.

What is clearly required is thus some transformed notion of responsibility beyond the simple multiplication of roles or the expansion of occupational role responsibility to encompass public safety, health, and welfare. Indeed, techno-scientific applications can remain ethically problematic even in cases where scientists and engineers have the best possible intentions and users have no conscious intention to misuse or abuse. (Think of the example of automobile pollution.) This situation constitutes the major ethical challenge we face today.

How are we to address the problematic consequences of collective actions? Technological risks are examples of special concern. The nature of many technological risks is far beyond the framework of individual responsibility. Such risks arise, as Charles Perrow (1984) has argued, as a consequence of an interaction of semi-independent systems, many of which may themselves be in part so complex as to be outside direct control. (Think of the examples of the economy or the legal system as well as those of the various sciences and fields of engineering.) Such risks often cannot even be constrained within the dimensions of some particular time and place, which makes the identification of possible victims impossible. For such risks it is thus not even possible to take out insurance. Many of the technological risks in our society have the same status as natural catastrophes (see, e.g., the argument of Beck, 1986).

In response to this problem, we propose what we will term an ethics of collective co-responsibility. The itemized inadequacies of occupational role point precisely in this direction. Such a collective ethics of co-responsibility arises from reflection on the social processes in which technological decision making is embedded. (It may even be interpreted as involving a renewed appreciation of Cicero's four-fold root of role responsibility.) That is, any new ethics must deal with the same substance as the old role responsibility ethics, namely with values and norms that restrict or delimit human action and thus enable or guide traditional decision making; but in the new ethics these values and norms will arise not simply in relation to occupational roles. How such an ethics of collective co-responsibility might work in the context of biotechnology has previously been outlined (Schomberg, 1998). Here it is appropriate to limit ourselves to indicating three general features and requirements for the implementation of such an ethics.

1. Public debate: *To be co-responsible includes being personally responsive.* It is clear that the norms of specific technical professions are insufficient because they arise from restricted perspectives. A true ethics of co-responsibility must be both interdisciplinary and even inter-cultural, in order to provide a standard of justice for evaluating and balancing conflicting occupational role responsibilities. If we fail to

provide such an ethics, we inevitably continue to aggravate the clash of cultures and unarticulated hostile responses to particular (globalized) technologies.

According to our view, an ethics of collective co-responsibility is expressed at the level of free (international) public debate in which all should participate. It is unethical and even unreasonable to make any one individual responsible for the consequences and/or (adverse) side effects of our collective (especially technological) actions. It is, however, ethical and reasonable to require individual participation in public debates (subject, of course, to the particular situation), or at least make this the default position for which persons must give reasons for being excused from such a duty. Upon everyone's shoulders rests a particular moral obligation to engage in the collective debate that shapes the context for collective decision making. It is not just engineers who do social experimentation; in some sense all human beings are engineers insofar as they are caught up in and committed to the modern project.

If we trace, for instance, the history of environmental challenges, we see that many issues which depend on the involvement of personally responsible professionals were first identified and articulated within the public sphere. Public deliberation does not primarily aim at creating of itself a reasonable consensus, but serves, among others, the function of presenting different relevant isssues to the more or less autonomous systems and subsystems of society – that is, to politics, law, science, etc. The typically independent discourses of politics, law, science, etc. are called upon to respond to issues raised in public debate. An appropriate response by the appropriate subsystem to publicly identified and articulated issues constitutes a successful socio-ethical response. Conversely, responsible representatives of the subsystems are drivers for new debates, when they publicize particular aspects of an issue that cannot be fruitfully resolved within the limits of some specialized discourse. The continuous interaction between the autonomous subsystem discourses and a critically aware public provides an antidote for frozen societal contradictions between opposing interests, stakeholders, or cultural prejudices.

2. Technology assessment: *To be collectively co-responsible involves developing transpersonal assessment mechanisms.* Although the institution of the public realm and interactions with the professionalized subsystems makes it possible for individuals to be co-responsive, these deliberations are in many cases insufficiently specific for resolving the challenges with which technological developments confront us – that is, they do not always lead to the implementation of sufficiently robust national or international policies. Therefore all kinds of specific deliberative procedures – for instance deliberative technology assessment procedures[19] – must be established to complement general public debate and to provide an interface between a particular subsystem

and the political decision-making process. The widely discussed consensus-conferences are one example of an interface between science and politics (see Mayer, 1997; Mayer & Geurts, 1998).

The implementation of ethics codes by corporations also constitute an interface between the economic sector, science, and stakeholder interest groups, while national ethics committees are often meant as a intermediaries between the legal and political system. Experiments with such boundary activities or associations have been, depending on the case, more or less successful.[20] They represent important experiments for enabling citizens to act as co-responsible agents in the context of technological decision making. Yet the absence of adequately deliberative forums is certainly one reason why we are not yet able to democratically plan our technological developments.

3. Constitutional change: *Collective co-responsibility may eventually entail constitutional or structural political change.* The initiation of specifically new forms of public debate and the development of transpersonal science and technology assessment processes may eventually require constitutional adjustments. Indeed, the adaptation of specific deliberative principles in our constitutions must not be ruled out.

Consider, for instance, the possible implementation of the precautionary principle, which is inscribed in the European Treaty and now also guides important international environmental deliberations (the Kyoto Protocol on Climate Change, the Biosafety Protocol, etc.). This principle lowers the threshold at which governments may take action to restrict scientific or technological innovation. According to the precautionary principle, one should err on the side of precaution. It does not have to be shown with certainty that bad consequences will follow; it is sufficient if there is an absence of scientific certainty in cases where there is some indication of possible serious or irreversible harm to human health or the environment. The very implementation of such a principle requires new and badly needed intermediate deliberative science-policy structures. It imposes an obligation to continue to seek scientific evidence and enables also an ongoing interaction with the public on the acceptability of the plausible adverse effects and the chosen level of protection.[21] The principle also requires companies to become more proactive and necessarily shapes their technoscientific research programs in specific ways.

CONCLUSION

The long history of role responsibility ethics as a mediation between ethical systems and social structures was progressively abandoned by modern philosophy in favor of an emphasis on principle responsibility ethics. A kind of empirical witness to the inadequacy of this shift can be found in the grassroots and largely non-philosophical efforts to revive role responsibility in engineering ethics. Yet attempts to enlarge the notion of role responsibility in technical professions such as engineering – from company loyalty to protection of public safety, health, and welfare – together with technosocial transformations in the nature of roles themselves, point toward needs for the development in advanced technoscientific societies of something beyond role responsibility. We have ventured to name this new attempt to mediate between ethical systems and social structures collective co-responsibility ethics, and to indicate what we take to be some of its central features.

NOTES

1. The history of this emergence, which is distinct from the emergence of engineering ethics itself, has yet to be written. Suffice it to note here that in the United States engineers first reached out to philosophers and philosophers first began to work with engineers in the 1970s. The single most representative result is Mike Martin (philosopher) and Roland Schinzinger (engineer), 1983, now in its 3rd edition, 1996. For some historical notes, see Mitcham 1997; and for some history of the applied ethics movement in general, including engineering ethics, Davis 1999.

2. See also Turner 1968. Banton 1999, p. 749, gives virtually the same definition: "A role is the expected behavior associated with a social position." A slightly more abstract definition is offered by Downie 1971, p. 128: A role is "a cluster of rights and duties with some sort of social function." Downie also references Emmet 1966.

3. Cicero's position certainly influenced and is reflected in Epictetus. See, e.g., Arrian, *Discourses of Epictetus* II, x.

4. This phrase, although commonly used to describe Christian existence, seems to derive from William Hazlitt (1778–1830). See Hazlitt 1821, Essay 10, "On Living to One's-Self," paragraph 3: "What I mean by living to one's-self is living in the world, as in it, but not of it."

5. George Herbert Mead (1863–1931) is particularly important for extending the notion of role beyond ethics by using it both as a concept to explain behavior in sociology and as a concept to explain the development of consciousness and knowledge in psychology. People are actors with roles (sociology). But people also have the imaginative ability of "taking the role of the other," which is the basis of self-awareness and human sociality. See, e.g., Mead 1934.

6. Neither the *Encyclopedia of Philosophy*, 8 vols. (New York: Macmillan, 1968) and *Supplement* (New York: Simon & Schuster, 1997), nor the *Routledge Encyclopedia of Philosophy*, 10 vols. (New York: Routledge, 1998) – indeed, not even the *Encyclopedia of Ethics*, 2 vols. (New York: Garland, 1992) nor the *Encyclopedia of Applied Ethics*, 4 vols. (San Diego: Academic Press, 1998) – have entries on role or role responsibility ethics. *The Routledge Encyclopedia* entry on "Responsibilities of Scientists and Intellectuals" (Montefiore 1998) assumes but does not analyze role responsibility; the

Encyclopedia of Applied Ethics entry on "Legal Ethics, Overview" (Tur, 1998) includes an analysis of the lawyer's role. Bayles 1981 and Koehn 1994 consider role responsibility but mostly to reject it in favor of principle and contract for the public good. The only extended philosophical analyses of role responsibility in English in the last half century are Emmet 1966 and Downie 1971; and Downie develops Emmet. This is in marked contrast with the literature in social theory, where Robert K. Merton, Talcott Parsons, and others have written at length on the topic of roles, although more as an explanatory than as a normative category. With regard to one specific role that has received special philosophical attention, see also Luban 1988, especially chap. 6, "The Problem of Role Morality," and chap. 7, "The Structure of Role Morality," pp. 104–147.

7. See also MacIntyre's pointed argument in "What Has Ethics to Learn from Medical Ethics?," *Philosophic Exchange* 37, that "any rational moral evaluation requires seeing an actor in some role . . . no one is ever an abstract moral agent . . . moral agency is embodied in roles such as that of the physician, the patient or the nurse." Thus, "If moral agency is exercised through roles, then the questions that ought to be addressed are much more specific than those with which moral philosophy is conventionally concerned . . . The moral agent turns out to be no more and no less than both the sum and the unity of his roles embodied in a single person. The abstract ghost of conventional ethics, man as such . . . has to be replaced by this much more interesting figure."

8. The best history of this phase two development in the United States is Layton 1971. Layton's analytic narrative goes up to World War II, and thus includes considerable material related to phase three as well. It is also worth noting that this same historical period witnessed formation of the first medical and legal associations in the United States. The American Medical Association (AMA) was founded in 1847, the American Bar Association (ABA) in 1878. But in contrast to engineering societies, the AMA from the outset adopted an explicit code of professional conduct drawn from the British physician Thomas Percival's influential book *Medical Ethics* (1803).

9. It was during this same period, in 1908, that the ABA also adopted a code of professional conduct. However, unlike the codes of the engineering profession, and as with the AMA code adopted earlier, it did so with reference to precedents and influences that extended back into the 1800s.

10. The idea that company loyalty in engineering is in truth a two-way relationship does not seem to have been argued. Cf. Solomon, 1991, p. 363: "It cannot be overemphasized that 'loyalty' here is a two-way concern; the employee may by virtue of his or her employment have special obligations to the company but the company has its obligations to the employee in turn."

11. The implicit argument here is that some roles are such that one cannot have any description of them that does not include a normative component. Bernard Williams 1972, p. 52, gives a succinct example: "[T]here is no notion of 'bank clerk' which does not involve a reference to the responsibilities, and the term refers to a role which can only be explained in relation to social institutions which give someone with that role certain functions and duties".

12. It should be noted, however, that rather than simply throwing out the paramountcy clause, Broome 1989, p. 9, simply proposes its reformulation as "The engineer shall not claim that engineering necessarily assures good public health and welfare. Instead, he shall advise the public of the risks associated with his work and seek to obtain public acceptance of these risks." Such a reformulation might make the principle slightly more enforceable – but only very slightly.

13. Although Martin and Schinzinger first advanced this thesis in 1983, subsequent editions of their textbook in 1989 and 1996 have neglected to draw any radical implications from this quite radical description of engineering.

14. See especially the analysis of Weber's theory of rationalization in Habermas 1984, chap. 2, and Habermas 1987, chap. 8, for overviews of these transformations.

15. This is almost certainly the most common comment on role responsibility ethics. Dorothy Emmet 1968, for example, observes in passing how individuals are often called upon "to meet the demands of a number of different and perhaps conflicting roles" and cites Barnard 1938 for documentary evidence. For other examples, see Downie 1971. Richard Wasserstrom 1983 makes the tension one between particular role responsibility and our obligations as rational moral agents to adopt universal moral principles.

16. The ethics of science has been much more institutionalized, but at the same time remained relatively internalist. See, e.g., Kristin S. Shrader-Frechette 1995.

17. One prominent synthesis of such approaches has become known as the Science, Technology, and Society (STS) movement. For a good introduction to STS, see Cutcliffe 2000; and Cutcliffe and Mitcham 2001.

18. In fact, Eatherly was not the pilot of the plane that dropped the Hiroshima atomic bomb on August 6, 1945. The pilot of that plane, the *Enola Gay*, was Paul Tibbets, and his bombardier Thomas Ferebee, neither of whom ever expressed any regret about his role. Eatherly was the pilot of a reconnaissance plane that preceded the *Enola Gay* and gave the go ahead. Questions have also been raised about the reality of Eatherly's feelings, but he still stands a symbol for a certain kind of response.

19. See, for instance, the contributions in von Schomberg, ed. 1999, especially Brekke & Eriksen 1999.

20. See David Guston 1999 for an extended argument in favor of what he calls boundary organizations as intermediaries between science and politics.

21. For further discussion, see again the six papers collected in von Schomberg (Ed.), 1999.

REFERENCES

Akin, W. E. (1977). *Technocracy and the American Dream: The Technocrat Movement, 1900–1941.* Berkeley, CA: University of California Press.

Anders, G. (1980). *Die Antiquiertheit des Menschen.* Vol. 1: *Über die Seele im Zeitalter der zweiten industriellen Revolution.* Vol. 2: *Uber die Zerstörung des Lebens im Zeitalter der dritten industriellen Revolution.* Munich: Beck.

Arendt, H. (1963). *Eichmann in Jerusalem: A Report on the Banality of Evil.* New York: Viking.

Banton, M. (1999). .Role. In: A. Kuper & J. Kuper (Eds). *The Social Science Encyclopedia.* (2nd ed.) (p. 749). New York: Routledge.

Barnard, C. I. (1938). *The Functions of the Executive.* Cambridge, MA: Harvard University Press.

Bayles, M. D. (1981). *Professional Ethics,* 2nd ed. Belmont, CA: Wadsworth. (2nd ed.), 1989.

Beck, U. (1986). *Riskogesellschaft: Auf dem Weg in eine andere Moderne.* Frankfurt am Main: Suhrkamp.

Bradley, F. H. (1911). *Ethical Studies.* New York: Stechert.

Brekke, O. A., & Eriksen, E. O. (1999). Technology Assessment in a Deliberative Perspective. In: R. von Schomberg (Ed.). *Democratizing Technology: Theory and Practice of a Deliberative*

Technology Policy. Hengelo, The Netherlands: International Centre for Human and Public Affairs (pp. 93–119).

Broome Jr., & Taft H. (1989). Can Engineers Hold Public Interests Paramount? *Research in Philosophy and Technology, 9*, 3–11.

Cutcliffe, S. H. (2000). *Ideas, Machines, and Values: An Introduction to STS Studies*. Lanham, MD: Rowman and Littlefield.

Cutcliffe, S. H., & Mitcham, C. (Eds.). (2000). *Visions of STS: Conterpoints in Science, Technology, and Society Studies*. Albany, NY: State University of New York Press.

Davis, M. (1999). The Ethics Boom, Philosophy, and the University. *Ethics and the University*. (ch. 1, 3–21). New York: Routledge.

Downie, R. S. (1971). *Roles and Values: An Introduction to Social Ethics*. London: Methuen.

Eatherly, C., & Anders, G. (1961). *Burning Conscience*. New York: Monthly Review Press.

Emmet, D. (1966). *Rules, Roles and Relations*. New York: St. Martin's Press.

Emmet, D. (1968). Ethical Systems and Social Structures. *International Encyclopedia of the Social Sciences*. New York: Macmillan Free Press.

Guston, D. H. (1999). *Between Politics and Science: Assuring the Integrity and Productivity of Research*. New York: Cambridge University Press.

Habermas, J. (1984). *The Theory of Communicative Action*. Vol. 1: *Reason and the Rationalization of Society*. Trans. Thomas McCarthy. Boston: Beacon Press.

Habermas, J. (1987). *The Theory of Communicative Action*. Vol. 2: *Lifeworld and System: A Critique of Functionalist Reason*. Boston: Beacon Press.

Hazlitt, W. (1821). *Table Talk*. London: G. Bell.

Ivan I. et al. (1977). *Disabling Professions*. New York: Marion Boyars.

Jonas, H. (1984). *The Imperative of Responsibility: In Search of an Ethics for the Technological Age*. Trans. Hans Jonas and David Herr. Chicago: University of Chicago Press.

Koehn, D. (1994). *The Ground of Professional Ethics*. New York: Routledge.

Layton, E. T. Jr. (1971). *The Revolt on the Engineers: Social Responsibility and the American Engineering Profession*. Baltimore: Johns Hopkins University Press. (2nd ed.), 1986.

Luban, D. (Ed.). (1983). *The Good Lawyer: Lawyers' Roles and Lawyers' Ethics*. Totowa, NJ: Rowman and Allenheld.

Luban, D. (1988). *Lawyers and Justice: An Ethical Study* (pp. 104–147). Princeton, NJ: Princeton University Press.

MacIntyre, A. (1966). *A Short History of Ethics*. New York: Macmillan.

MacIntyre, A. (1984). *After Virtue: A Study in Moral Theory*. (2nd ed.). Notre Dame, IN: University of Notre Dame Press.

Martin, M. W., & Schinzinger, R. (1983). *Ethics in Engineering*. New York: McGraw-Hill. (2nd ed.), 1989. (3rd ed.), 1996.

Mayer, I. (1997). *Debating Technologies: A Methodological Contribution to the Design and Evaluation of Participatory Policy Analysis*. Tilburg, The Netherlands: Tilburg University Press.

Mayer, I., & Geurts, J. (1998). Consensus Conferences as Participatory Policy Analysis: A Methodological Contribution to the Social Management of Technology. In: P. Wheale, R. von Schomberg, & P. Glasner (Eds), *The Social Management of Genetic Engineering* (pp. 279–301). Aldershot, England: Ashgate.

Mead, G. H. (1934). *Mind, Self, and Society: From the Standpoint of a Social Behaviorist*. C. Morris, (Ed.). Chicago: University of Chicago Press.

Mitcham, C. (1987). Schools for Whistle Blowers. *Commonweal, 114*(7), (April 10), 201–205.

Mitcham, C. (1997). Engineering Ethics in Historical Perspective and as an Imperative in Design. *Thinking Ethics in Technology* (chap. 8, pp. 123–151). Golden, CO: Colorado School of Mines.

Montefiore, A. (1998). Responsibilities of Scientists and Intellectuals. *Routledge Encyclopedia of Philosophy.* New York: Routledge, *8*, 287–290.

Perrow, C. (1984). *Normal Accidents: Living with High-Risk Technologies.* New York: Basic Books. Revised edition, Princteon, NJ: Princeton University Press, 1999.

Price, D. J. de Solla. (1963). *Little Science, Big Science.* New York: Columbia University Press.

Percival, T. (1803). *Medical Ethics.* Manchester: Nicholson.

Sarbin, T. R. (1968). Role: Psychological Aspects. *International Encyclopedia of the Social Sciences.* New York: Free Press, *13*, 546–552.

Schomberg, R. von. (1998). Democratizing the Policy Process for the Environmental Release of Genetically Engineered Organisms. In: P. Wheale, R. von Schomberg, & P. Glasner (Eds), *The Social Management of Genetic Engineering* (pp. 237–248). Aldershot, England: Ashgate.

Schomberg, R. von (Ed.). (1999). *Democratising Technology: Theory and Practice of a Deliberative Technology Policy.* Hengelo, Netherlands: International Centre for Human and Public Affairs.

Shrader-Frechette, K. S. (1995). *Scientific Research Ethics.* Lanham, MD: Rowman and Littlefield.

Solomon, R. C. (1991). Business Ethics. In: P. Singer, (Ed.), *A Companion to Ethics* (pp. 354–365). Cambridge, MA: Blackwell.

Tur, R. H. S. (1998). Legal Ethics, Overview. *Encyclopedia of Applied Ethics.* Boston, Kluwer, *3*, 59–70.

Turner, R. H. (1968). Role: Sociological Aspects. *International Encyclopedia of the Social Sciences.* New York: Free Press, *13*, 552–556.

Vaughan, D. (1996). *The Challenger Launch Decision: Risky Technology, Culture, and Deviance at NASA.* Chicago: University of Chicago Press.

Wasserstrom, R. (1983). Roles and Morality. In: D. Luban (Ed.), *The Good Lawyer: Lawyers' Roles and Lawyers' Ethics* (pp. 25–37). Totowa, NJ: Rowman and Allanheld.

Williams, B. (1972). Morality: *An Introduction to Ethics.* New York: Harper and Row.

A DIALOGUE ON ENGINEERING DESIGN AND LAW

CODES OF CONDUCT, THE LAW, AND TECHNOLOGICAL DESIGN AND DEVELOPMENT

Henk Zandvoort

Starting from a discussion of legal and other obstacles to the socially responsible behavior of engineers and of corporations employing engineers, I explore how the legal system can promote the socially responsible behavior of engineers and of corporations. In the first two sections I develop the following two claims:

(a) If the right of everyone to be informed about risks caused by others is to be protected, then conflicts that exist between the legal systems and the obligation of engineers to inform the public should be removed (Section 1).
(b) If behavior is to be promoted that is consistent with sustainability and the responsible management of technological risks, then the legal systems should be based on strict liability (Section 2).

I conclude with an exploration of the possible impact on technological design and development of stricter forms of liability, especially with respect to safety, health, environment, and sustainability (Section 3).

The Empirical Turn in the Philosophy of Technology, Volume 20, pages 193–205.

ISBN: 0-7623-0755-2

1. CODE OF ENGINEERING ETHICS AND THE RIGHT TO BE INFORMED

Duty to Inform

Engineers who want to do their work in accordance with the codes of conduct framed by their professional societies meet with severe legal obstacles (see, e.g., Unger, 1994 for a general picture). Here we are concerned with one particular obstacle. It relates to the following obligation, contained in some form or another in many codes of engineering ethics:

(1) Engineers shall inform the public about the possible consequences of their work.

An employee who wants to live up to (1) is risking a conflict, both with his employer and with the law. The following examples from Dutch law make this clear. Similar conflicts arise in other legal systems.

Labor contract law forbids employees to disclose facts concerning their employer's business that the employee ought to keep secret. Violation is a legal ground for dismissal. The law does not specify which information ought to be kept secret. (A connection with, e.g., codes of ethics, is absent.) Hence, in case of a lawsuit the judge must decide on the interpretation of the law. But judicial decisions of this type tend to weigh business interests heavily against the values that the codes of conduct aim to protect. This can be explained from the fact that as a rule, only interests are taken into account that can be associated with a direct financial valuation.

The penal code prohibits the disclosure of business information on which secrecy was imposed. It is true that there is a provision that a person who "could assume in good faith that the general interest required disclosure" will not be punished. This statement is not further specified in the code, and judicial decisions show that it offers very little solace for employees who want to live up to (1). As a consequence, persons living up to (1) are risking grave personal difficulties and disadvantages.

Right to be Informed

The principle expressed by (1) is not a value in itself, but stands in the service of the following:

(2) All people should be informed about risks created by others that may affect them.

Because of the described conflict, living up to (1) will often require martyr-like behavior. Given that such behavior is scarce, society is ill-advised to rely on (1) for the protection or implementation of (2), as long as the conflicts between (1) and the legal system continue to exist. Hence, if obeying (1) is considered important because (2) is considered important, then the legal system should be adapted so as to make obeying (this aspect of) codes of conduct practically feasible.[1] The required adjustments of the legal system are not simple, since they may affect other legal rights, such as the right to protection of business secrets. The situation calls for substantial contributions by legal theorists.

Western constitutions do not explicitly recognize (2), and the contemporary legal systems do not support (2). As an example, take the situation with respect to public administration. In the Netherlands, the availability to the public of information used in the public administration processes is governed by the *Wet Openbaarheid Bestuur* (Law [concerning] Open Government). Important elements of this law are that: (i) information is only provided upon request; (ii) in many cases requests can be and are refused because the information requested falls under exclusionary clauses; (iii) if answers to requests are vague, incomplete, or not given at all, there are few possibilities to respond. Because of these elements, the *Wet Openbaarheid Bestuur* may in practice severely conflict with principle (2).

Probably the entire subject of the right to be informed about risks caused by others was not of practical importance at the time the Western constitutions were framed. But after two centuries of technological development we are in an entirely different situation. Freedom of speech and expression was included in the constitutions, and might to a degree stand in the service of (2). But as the examples show, this right is essentially limited for employees.

2. CODES OF CORPORATE CONDUCT AND LIABILITY

The question to be pursued in this section is: How can the socially responsible behavior of corporations best be brought about? As in the previous section, I start with a consideration of codes of conduct. After having established and explained their limited effectiveness, I will discuss the actual and potential role of legal liability.

Codes of Corporate Conduct

Many corporations have adopted codes of conduct concerning safety, health, environment, sustainability, and human rights. Such codes express restrictions on a corporation's behavior that go beyond what is legally permitted or required, or that fill up gaps where laws are absent or unclear. At best, these codes might function as self imposed restrictions on a corporation's behavior that otherwise is guided by the market principle.

Despite the adoption of such codes, sustainability is not in sight, and technological risks are not under control. Witness the greenhouse effect, ozone depletion, chemical substances diffusing into the environment, and genetic engineering in agriculture. Apparently, codes of corporate conduct are not very effective in bringing about the desired behavior.

The limited effectiveness of codes of corporate conduct should not surprise, given the problems that attach to such codes. They are often stated in vague and subjective terms, such that it is unclear what would be an infringement. For instance, if it is stated that health, safety, environment, human rights, etc., "*will be taken into account*", or "will be taken into account *as much as possible*", what do these phrases mean? In addition, systematic and independent control of their actual observance is often lacking, and infringements are not met with sanctions. Hence, a corporation that follows a more stringent code, or that more stringently follows a shared code, tends to give itself a competitive disadvantage. For these reasons it is a naive presumption that codes of corporate conduct can effectively restrict the profit-principle. The latter is not only more clear, but is also decisive for the corporation's continued existence.

A final criticism concerns the *procedures* by which codes of conduct are framed. We may presume that codes of conduct are adopted in order to establish or maintain a favorable reputation with clients or the public. As in other matters, a corporation will weigh its costs from adopting a code against its expected benefits. But there is no guarantee that a code that originates from such considerations is optimal in a broader sense as well. Corporate codes may also result from deals with governments or pressure groups. But such a mechanism cannot be considered optimal from the point of view of rational collective decision making.

The above does not imply that codes of corporate conduct are meaningless. Their existence expresses the broad recognition that existing laws are inadequate to bring about the desired socially responsible behavior. But from this it does not follow that the codes themselves can compensate for an inadequate legal system.

Prisoner's Dilemmas

To get a better understanding of the general problem and its solution, consider a situation where the following holds:

(a) Each would be better off, at least in the long run, if all would restrict their behavior in certain ways.
(b) Individual compliance with the restrictions results in a considerable disadvantage compared to non-compliers, whereas the negative overall effect of one or a few non-compliers is relatively small.

Such situations, called prisoner's dilemma situations, have been studied in the theory of games (see, e.g., Morrow, 1994, and references therein). They resemble the situation in which corporations (together with other autonomous actors) find themselves with respect to sustainability and risks. In prisoner's dilemma situations, it is often attractive for an individual not to comply with the restrictions, even if that individual is convinced of (a). It thus is possible that no one complies, although all are convinced of (a). A solution to such situations is the introduction of common prescriptions, and the appointment of an arbiter who controls the observance of the rules and who confronts infringement with sanctions. This solution, known for four thousand years, comes down to the introduction an adequate legal system.

Self-Determination and Liability

Hence the next question is what an adequate legal system should look like, if the problem is how to bring about sustainability and acceptable risks. Here a consideration of liability is crucial. Liability is a legal operationalization of responsibility. When people are legally liable for certain actions or behaviors this means that they can be forced to repair or to compensate for damages caused.[2] Below, I introduce a concept of liability that does not correspond to actually existing legal concepts of liability. This defined concept of strict liability is introduced, because it can be derived from ethical principles, and because it can be used to assess actually existing legal forms of liability.

I will call liability *strict* if all persons are liable for the possible consequences of their actions, unless all who may experience these consequences have given their informed consent. This informed consent may be given in either of two ways:

(1) By means of a general rule (e.g., a law on which there is consensus); or
(2) From case to case.

Strict liability as defined here has its basis in ethics (Van Velsen, 2000). It is complementary to the principle of self-determination, saying that all individuals are free to do what they please as long as they do not harm others.[3] An implication is that actions are right if and only if: either there are no (possible) consequences for others; or those who will experience the (possible) consequences have consented after having been fully informed.

Strict liability promotes prudence, and therefore promotes sustainability and control of risks. Therefore, the principle of strict liability may be valued not merely as a consequence of ethical principles, but also as providing a sound basis for a legal system that should promote sustainability and the control of risks.

Liability and the Legal Systems

In the legal systems as they actually exist, liability is almost never strict in our sense, but subject to conditions. An example is "fault liability". But also product liability, although often called "strict", is in essential respects not strict in our sense, at least not according to Dutch law. Thus, product liability (under Dutch law) holds producers liable for damages caused by defects in their products, *unless* "it was not possible to discover the existence of the fault, given the state of scientific and technological knowledge at the time the product was brought in circulation"; in this case the producer is *not* liable.

In an innovative technological society governed by conditional liability, more and more activities come into being with risks or side effects for which the actors are not held liable. Usually, the advantages of new technological activities and possibilities are clear from the outset, whereas important harmful effects become manifest only later. Damages remain with affected non-actors, although they were not able to influence the development and production of the technologies in question. Consider biotechnology. On a large scale, genetically engineered organisms are being introduced. Who will repair or compensate if negative effects materialize that opponents are warning against, or that are yet unforeseen?

Contemporary law deviates from strict liability as defined above in the following two important respects:

(1) The relevant laws (such as the laws governing new technological activities) often do not have the consent of all who may experience the possible consequences.
(2) Violation of the laws does not always lead to the obligation to repair or compensate.

These are related points. The willingness to give one's consent to certain activities (such as the introduction of genetically engineered organisms) will increase as repair of or compensation for damage, if any, is better guaranteed.

Legal history shows that, until into the 19th century, private law was more than now based on strict liability (Van Dunné, 1993; Horwitz, 1977). This holds for both aspects distinguished above. More than at present there was consensus on the then existing legal rules, such as in the domains of property and contract. And violations were met with liability that was much less than at present subject to conditions. The doctrine of "no liability without fault" only emerged in the course of the 19th century, and limited liability of corporations was still a marginal phenomenon. The 19th and 20th centuries subsequently saw the spread of non-consensual law in areas such as tort, expropriation, and corporate law, together with the introduction of conditional liability such as fault and limited liability. These transformations were driven by the desire to promote technological development (Horwitz, 1977; 1992). Because of the relative absence of clearly harmful side effects, these transformations may have seemed acceptable at that time. But after more than a century under its regime this is different. This is notwithstanding the fact that in the 20th century there have been, in some domains such as product liability, partial moves toward stricter liability.

To conclude, conditional liability, introduced in the 19th century and persisting in our time, is an important cause of the present situation of non-sustainability and uncontrolled risks. In contrast, strict liability promotes prudence, and hence promotes sustainability and control of risks. Strict liability may thus be preferred not merely on abstract grounds of ethics and justice, but also from considerations of responsible management and control of technology.[4]

At this point I want to return to prisoner's dilemmas. As was stated, corporations are in a prisoner's dilemma with respect to aspects such as sustainability and risks. As long as this situation continues, organizations that live up to codes of conduct are risking competitive disadvantages. This explains the hesitations and reservations of corporations and of organizations in general with respect to the entire subject of codes of conduct. But it should be clear that the main problem of organizations is not the existence of boundary conditions to their activities, but the existence of *non-uniform* boundary conditions. For only the latter create inequalities with respect to competition.

Responsibility of Corporations and of Their Employees

The above considerations on responsibility and liability not merely pertain to corporations as artificial legal persons, but to all free and autonomous actors. Hence they would apply to engineers in the performance of their professional activities if they could be considered free. But at present the latter is not a valid assumption. For labor contract law makes them agents, essentially taking orders from their employers. Thus, in Dutch law an essential element of a labor contract is that the employer has authority over the employee.

Consistent with this, the law of corporations holds the corporation responsible (and liable) for whatever its employees, including engineers, do or fail to do in the course of their work. Exceptions occur only in cases such as when employees were grossly negligent or deliberately caused damage. Hence, although it would be consistent to extend the above analysis of responsibility and liability to individual engineers (and to other individuals related to a corporation), the results would require drastic changes in the (laws governing) responsibility in organizations. This issue will not be pursued here.[5] It should be noted, though, that the introduction of strict liability for corporations (and not for employees) would certainly make it easier for engineers to live up to the various aspects of their professional codes.

3. IMPLICATIONS FOR RESEARCH AND DEVELOPMENT IN SCIENCE AND TECHNOLOGY

In what ways might the legal development of stricter forms of liability with respect to subjects such as safety, health, environment, and sustainability influence scientific research and technological design and development? This topic will be pursued in the present section.

More precisely, I will discuss the influence of regulation and liability taken together, where liability is now taken in the ordinary legal sense of being obliged to repair or compensate for the negative consequences of certain acts. As is clear from Section 2 above, liability and regulation are related issues. For instance, if the liability for certain activities becomes more strict, that is, less subject to conditions such as actor's "fault" and the "forseeability" of damage or harmful side effects, the potentially harmed will have less need for regulation that would keep potentially injurious or dangerous behavior within bounds. In the extreme (but also rare) case that possible damages from actions can be completely repaired, one could in principle do without regulation altogether. Rather than making predictions, my aim is to explore possible implications of stricter liability/regulation.

3.1. Implications for the Employment of Scientists and Engineers

Following the conclusion in Section 2 above, one may expect better chances for the realization of sustainability and acceptable levels of risk if stricter forms of liability would be introduced. That is, better performance on matters that, under headings such as safety, health, environment, and sustainability, belong to the central concerns stated in codes of conduct both for engineers and for corporations.

In addition, under stricter forms of liability, more research and development might go into each new technological product or process. That is to say that, other things remaining equal, a greater percentage of the price of a technological product or service might be made up of costs for R&D. This means that the relative contribution of scientists and engineers to the added value in technological products or processes would increase. The time required to develop new products or processes might increase as well. Also the scale of development and production activities might increase, which might take the form of mergers between existing industries.

The combined phenomena of increasing the share of R&D in value added, increasing time needed to bring products to the market, and merging of companies, can be observed in the pharmaceutical industry. I do not claim that in this particular case these phenomena have been brought about, or exclusively brought about, by more stringent regulation and liability. Nevertheless the example is relevant to the topic addressed here. For in this industry neither the high contributions of R&D to value added, nor the long time needed to develop new products, have stood in the way of profitability and growth.

Another relevant question is this: To what extent are people willing and able to pay for new technology performing functions in a more sustainable or less risky way? Here the drugs industry is also interesting. The contribution of R&D to the gross added value (called the R&D-intensity) in this sector is very much higher than in the industry as a whole. According to a recent report (Hollanders and Verspagen 1998), the average R&D-intensity of the pharmaceutical industry in 10 industrialized countries in 1994 was 28%, comparing to 6.5% for industry as a whole. Other industrial sectors with high R&D intensity were computers (26% average), electrical machinery (20%) and airplanes (27%). Sectors with relatively low R&D intensity were, e.g., textiles (1%), food (1½%), metals (3%), oil (4%), and shipbuilding (3%).

The considerable amounts of R&D that have gone into these specific technological products (pharmaceuticals, computers, airplanes) have actually been paid for. This is not to say, of course, that they have been *directly* paid by their users. Think of the role of government in financing health care. Think of the role of government support, either directly or indirectly through the tax system, in the airplane and airline business. It does show, however, that as a matter of fact there is a considerable buying force for R&D activities. That buying force might well be focused more than it is at present on high-tech products and processes (that is, with high R&D intensity) designed to perform functions in more sustainable or safer ways.

In addition, the pattern of overall economic expenditure in Western societies is such that there is ample space for an increased share of R&D-intensive

technology, and hence of R&D as such, in that overall expenditure. In fact, this space has been created by technological development itself, in the form of a continuously increasing economic efficiency in production. As a specific example, compare present R&D expenditures with the huge amounts spent on mere advertising.

An increased share of R&D in the Gross Domestic Products of countries is equivalent, of course, to more employment for scientists and engineers. It often seems that stakeholders of the scientific and technological communities are not aware of this, when they reason that increased liability would be detrimental to the employment of scientists and engineers.

Of course, money (and more to the point, working power) can be spent only once. Hence there is the issue of choice. At present, the actual choices are negatively influenced by existing prisoner's dilemma situations which can, at least in part, be solved by the introduction of stricter liability.

3.2. Implications for the R&D Methodology and for Responsible Exercise of Profession

Stricter liability forces engineers to "make things better": to make things in such a way that there will be fewer side effects on health, safety, environment, and sustainability. Hence, stricter liability may influence design procedures and methods, and the methodology of applied science. If things must be made better, the art of applied science and technology becomes more difficult and perhaps even more challenging than at present. Rather than removing the thrill from applied science and technology, as some seem to fear, stricter liability may actually contribute positively to that thrill.

Stricter liability leads to more consensus on technological contributions to society and human well-being. This is a necessary condition for being able to state in a non-subjective, non-arbitrary way that technology achieves progress.

Stricter forms of liability will also lead to more consistency between codes of conduct (both for engineers and for corporations) and self-interests (on all levels). This means that under stricter forms of liability the responsible exercise of the professions of scientists and engineers will be easier. The present defensive and often authoritarian attitudes of (stakeholders of) the scientific and technological communities toward society would disappear. Likewise, there would no longer be reason for the (implicit or explicit) complaints of many scientists and technologists, that society sees them primarily as unfeeling technocrats and as scapegoats, rather than as valuable members of the community who enhance the well-being of all.

The last two points (more consensus and more consistency with codes of conduct) come down to a reduction of ethical problems associated to technology and engineering.

For all reasons mentioned above, doing applied science or engineering may under stricter forms of liability give more satisfaction, and hence might make a career in science and engineering more attractive for talented young people than presently is the case.

3.3. International Competition and National Laws

It is a commonplace that knowledge has become an increasingly important factor in economic competitiveness, in addition to or in partial replacement of the traditional factors of labor and capital. Stricter national regulation and liability with respect to safety, health, environment, and sustainability challenges and stimulates the further development of this third factor, and hence may stimulate the future competitiveness of a country's industry.

In response to this argument, it is often remarked that a country that one-sidedly strengthens its legal regulations and liability with respect to safety, health, environment, and sustainability will lose its competitiveness to countries maintaining less strict laws. Such a country, it is said, would price itself out of the market. But this may not be true. In the longer run it may rather lead to competitive advantages, since the introduction of stricter regulation and liability in the other countries is just a matter of time.

It has been put forward by Ashford (1993) that, if one believes in the strategic value of stricter legislation, there is a danger in taking too small, incremental steps. Ashford was considering regulation in particular, but his argument is equally relevant to liability. In the case of small regulatory steps, Ashford argued, there are more opportunities to continue adapting existing technologies to new standards (e.g., by means of end-of-pipe adaptations), and less opportunities for developing new technologies, based on design principles that would enable to meet future legal standards more cost-effectively than what can be accomplished by adapting present technology.[6]

The topics discussed here may greatly benefit from comparative empirical research. A first desideratum is to have reliable and easily accessible information on the comparative stringency of legislation pertaining to safety, health, environment, and sustainability in various countries. Too often statements on this issue go without references to clear sources. Next, one may try to correlate technological innovations in a country with the nature of its legislation regarding safety, health, environment, and sustainability, and try to discover relations of cause and effect. It would be stimulating to see examples of successful technological innovation triggered or stimulated by advanced (in the sense discussed above) national legislation.

ACKNOWLEDGMENT

I thank J. F. C. van Velsen for his comments on a draft of the first two sections of this study.

NOTES

1. Space does not permit an exposition of the following claim. The right to be informed expressed by (2) is a prerequisite for technological progress in any sense that is consistent with the principle of self-determination to be mentioned in section 2 below.

2. In my response to Vlot's comments below I will distinguish between situations where repair is possible in principle, and situations where repair is impossible because of irreversible consequences and hence compensation is the remaining option. I will also address the possibility that actors may not always have sufficient means to repair or compensate for damage done.

3. This principle is contained as article 4 in the *Déclaration des droits de l'homme et du citoyen du 26 août 1789* of the French Revolution. The principle was defended by J. S. Mill in his essay *On Liberty* of 1859.

4. Ashford and Stone (1991, p. 417) have tried to assess the effects of liability for harm due to chemical production and products (excluding pharmaceuticals, and the disposal of chemical residues or wastes) as deployed under the (then) prevailing tort law system in the USA. Questions they wanted to address were: (1) Does tort liability for chemical harm promote safety and health? (2) Does tort liability for chemical harm stimulate or inhibit innovation? (3) In trying to control chemical hazards and foster technological advance, does tort liability for chemical harm impose an unacceptably high cost on consumers and industry? They estimated that the total liability costs of chemical firms due to acute hazards (such as accidents with chemical plants) were less, and for chronic hazards of chemicals far less, than the corresponding social costs. (It should be noted that an estimate of social costs requires a host of assumptions, which may not all be equally well founded.) They concluded that the prevailing liability laws provided insufficient deterrence (or "underdeterrence") for the chemical industry (to keep their hazardous activities within socially acceptable bounds). They reported evidence of liability-induced chemical innovations, especially in the area of process technology (triggered by occurrences such as Bhopal and Seveso). They also mentioned examples of process safety improvements such as the replacement of organic solvents with water based cleaners that were "probably stimulated more by regulatory prodding, particularly restrictions on the disposal of hazardous wastes, than by liability concerns." Their overall conclusion was that (under the prevailing tort laws) "liability plays only a minor role in stimulating the development of less chronically hazardous chemical products and processes." But they also expressed that "[the fact] that we have observed the development of both new products and processes when there is small or modest economic

deterrence leads us to believe that increasing the economic costs associated with liability by more fully internalising the social costs associated with chemical production would stimulate new [and safer, HZ] technologies."

5. I will address the issue in my response to Vlot's comments.

6. Ashford assumed also that existing industry prefers incremental legislation above non-incremental changes, because of the relative weight existing industry attaches to short term profitability considerations, and because non-incremental changes might stimulate competition from outside the vested industry. This is why he warned governments (and the electorate) not to let their policy for future legislation be influenced too much by the desires of existing industry.

REFERENCES

Ashford, N. A., & Stone, R. F. (1991). Liability, Innovation, and Safety in the Chemical Industry. In: R. Litan & P. Huber (Eds), *The Liability Maze: The Impact of Liability Law on Safety and Innovation* (pp. 367–442). Washington, DC: Brookings Institution.

Ashford, N. A. (1993). Understanding technological responses of industrial firms to environment problems: implications for government policy. In: N. A. Ashford & J. Schot (Eds), *Effective Government Action* (pp. 277–307). Washington, DC: Island Press.

Van Dunné, J. M. (1993). *Verbintenissenrecht, deel 2.* 2e herziene druk. Deventer: Kluwer.

Hollanders, H., & Verspagen, B. (1998). *De invloed van de sectorstructuur op de Nederlandse R&D uitgaven.* Rapport uitgevoerd in opdracht van het Ministerie van Economische Zaken. Maastricht: Merit.

Horwitz, M. J. (1977). *The transformation of American law 1780–1860.* Cambridge, MA: Harvard University Press, especially chapters 1–3.

Horwitz, M. J. (1992). *The transformation of American law 1870–1960.* New York and Oxford: Oxford University Press, especially chapter 3.

Morrow, J. D. (1994). Game theory for political scientists. Princeton, NJ: Princeton University Press.

Unger, S. H. (1994). *Controlling technology: Ethics and the Responsible Engineer* (2nd ed.). New York: John Wiley.

Van Velsen, J. F. C. (2000). Relativity, universality and peaceful coexistence. *Archiv für Rechts- und Sozialphilosophie, 86*(1), 88–108.

TOWARD A JURIDICAL TURN FOR THE ETHICS OF TECHNOLOGY? AN AEROSPACE CASE

Ad Vlot

INTRODUCTION

Henk Zandvoort accurately identifies legal obstacles to the responsible behavior of engineers. He makes two claims. According to him safety and sustainability in technological development can be promoted by: (1) the legal protection of engineers who disclose information on risks to the public, and (2) the introduction of strict liability which will stimulate prudence and hence sustainability and a more adequate control of risks. The aim of my chapter is not primarily a thorough philosophical analysis of these claims; I was asked to confront both claims with my personal struggle to be an ethically responsible human being as well as a professional aerospace engineer. I sympathize with Zandvoort's intentions and ideas. However, changes in the juridical system are necessary, but are not sufficient. I will try to make plausible that the "juridical turn" of Zandvoort as an operationalization of the ethical principle of self-determination remains external to some real engineering dilemmas within the contemporary structure of modern technological development.

The Empirical Turn in the Philosophy of Technology, Volume 20, pages 207–218.

ISBN: 0-7623-0755-2

1. WORLDVIEW AND THE STRUCTURE OF MODERN TECHNOLOGICAL DESIGN AND DEVELOPMENT

Every analysis of ethical problems is performed from a specific perspective that is shaped by a worldview, personal experience, and psychological character. The description of my case history will be expressed from the perspective of my worldview and background, which is related to Reformational philosophy, the Dutch Amsterdam school of philosophy founded by Herman Dooyeweerd during the first part of the 20th century. His philosophical framework basically consists of three parts (Dooyeweerd, 1969):

(1) an epistemological part in which he tries to prove that every philosophical position will be inevitably founded in presuppositions of a religious kind,
(2) a philosophical analysis of the "groundmotives" (the "thrust") of Western culture, and
(3) a philosophical analysis of the given structure of reality based on empirical "states of affairs" by which he is able to analyze entities by a fairly complex system of different aspects (modalities).

In his view these fundamental aspects of reality can be distinguished for every entity or thing. They are divided in the "lower" natural ones (mathematical, physical, biological, etc.) and "higher", cultural modalities (social, historical, linguistic, ethical, etc.). Fixed natural laws govern the natural aspects, but the cultural laws have a different character due to human freedom. This is a "state of affairs" that is reflected in the fundamental difference between the sciences and the humanities, which gives rise to existence of the "two cultures" (Snow, 1965) and to the reduction of engineering education to the exact sciences (Meijers, 1998).

Basically, technology uses and exploits natural aspects of objects in such a way that they can fulfil cultural functions. As an aerospace engineer I use mainly physical and mathematical laws to "create" aircraft and space vehicles that have primarily a function in the social realm (transport of people and communication) but also possess all the other functions, even a strong religious one (Noble, 1997). In this way materials found in nature (e.g., ores and wood) also receive an inherent ethical dimension in every technical artifact. In this view the ethical aspect of entities (things, actions, artifacts, etc.) is only one of the basic modalities of reality, and is fundamentally different from and cannot be fully reduced to the other aspects like the historical, the social, the aesthetical, the technical, or the juridical aspects. For example: ethical laws cannot be fully operationalized, controlled, or implemented like an algorithm in a computer code, which means that they cannot be fully treated in a technical way. However, the ethical aspect of things and actions is certainly related to all the other aspects. This means that an ethically sound design should obey social norms (e.g., the engineer should inform the public about risks), aesthetical norms (e.g., it must obey aesthetical rules of beauty and harmony), economical considerations (e.g., careful use of scarce resources), juridical laws (e.g., do not violate governmental regulations), etc. They also need the right *technical* implementation in legislation. This approach is called the "multi-modal" approach of ethics.

In my view there are at this moment two philosophical movements that tend to treat the ethics of technology from one isolated aspect: the *social constructivist* and the *juridical* movements. In social constructivism ethical problems are modeled and solved as a societal debate between different actors involved in a technology. The risk of this approach is relativism ("let a thousand lights shine"). From the juridical viewpoint, as expressed by Zandvoort, there is an isolated stress on the juridical operationalization of ethical principles. Both movements rightfully oppose the mainstream of a purely technicistic reduction in which ethical choices evaporate in engineering practice and the technical "one best way" is pursued.

Van Riessen and his successor Schuurman, two professors in philosophy at the Delft University of Technology, expanded Dooyeweerd's structural analysis of reality and culture (cf. Mitcham, 1994, p. 34). Both scientists were educated as engineers and were therefore able to perform a realistic philosophical analysis of the structure of technology. Van Riessen's dissertation of 1949 was written as a critique of the philosophers of his day who treated the role technology in society as a black box. In this respect he was was – in the terminology of Hans Achterhuis (1997) – not a "classical" but an "empirical" philosopher of technology. However, like classical philosophers such as Heidegger and Ellul, Van Riessen focused mainly on the essential way in which *modern* technology differed from *traditional* technology. But he tried to reveal the empirical basis of the critique of these classical philosophers of technology.

In Van Riessen's view modern technology has a typical structure in which artifacts are designed and produced from neutral functions, without meaning apart from the whole artifact. "Neutral" in this sense means isolated, non-specific, abstract, without meaning in itself, replaceable and normalized. These neutral parts and processes can be optimized in a quantitative way and are integrated during design and production by stepwise integration into an object. For example: materials such as aluminum alloys are quantified and documented in a database as design allowables by a qualification process. They become meaningless parts like rivets, sheets, and frames in the form of quantitative characteristics that may be used in an abstract design optimization process to design a structure and a manufacturing process. The neutral parts are assembled in a manufacturing process to a specific, complex structure. But these neutral parts, like sheets, amplifiers, and modules, can also be used for completely different artifacts.

Engineers are thus separated from the fabrication and the societal effects of the artifact. They prepare in their minds and on their screens in an abstract way the artifact and the production process. Engineers are also typical for modern technology. By using the described systematic approach of analysis and integration they have a theoretical "handle" to optimize the artifact. Their knowledge is laid down in and derived from the *technological sciences*. These sciences study and develop technology in a theoretical way and by laboratory experiments. Typical for modern technology is therefore also the technical university where engineers are educated and where technology is studied and developed in a scientific manner. One can say that modern technology is not applied physical sciences – in fact the relation between the fundamental sciences and technology is often not so intertwined – but that modern technology is primarily applied technological science. In the framework of the analysis of Dooyeweerd these technological sciences should not be part of the natural sciences such as physics and chemistry, but they are part of the humanities and the cultural sciences. Engineers do not create physical objects, but cultural (often social) ones.

In the traditional form of technology the same craftsperson is the designer and the maker, and the parts of a product are unique for that specific product. Standardized bolts, rivets, switches, transistors, filters, etc. are typical of modern technology. In this way the technological design method which isolates the different disciplines corresponding to the various neutral elements becomes a powerful tool which is used by engineers in combination with automation to control the designed artifact before it becomes operational. According to Van Riessen this systematic design method does not correspond with reality in which every thing is unique, variable and embedded in relationships. Therefore the art of engineering involves the "feeling of the engineer" for dealing with frictions between the result of theoretical optimization and reality, such as variations, wear, environmental influences and, above all, human freedom that may disturb the process or functioning of the artifacts.

Modern design is a highly systematic, analyzing process. Vincenti (1990) describes the different levels by which the design of the artifact in modern design methodology is broken down: from *project definition* (e.g., large passenger aircraft with 20% lower operational costs than its competitor), global design (e.g., double deck layout), *components* (e.g., fuselage panels), *disciplines* (e.g., material science, maintenance) to *specific problems* (e.g., durability of materials). This building block approach is followed for every neutralized function, component, and subsystem. Ethical considerations are mainly encountered during the project definition and maybe finally during the integration phase. In between, the design is reduced to the technical aspects in the various disciplines related to the neutral parts. The power of this method is that it allows the appointment of specialists who concentrate on one specific aspect of one neutral part, e.g., an aerodynamicist specialized in high lift devices of the aircraft wing, a material expert specialized in bonding processes, etc.

To summarize Van Riessen's views, I conclude that a specific design method laid down in the technological sciences is the foundation of modern technology with the following characteristics:

(1) A separation between production (manufacturing), operation (use), and preparation (design and management) by which ethical responsibilities and choices are shifted toward the preparation.
(2) The division of artifacts and processes into isolated "neutral" (also ethically neutral! – parts and functions.
(3) A drive to apply automated machines for a more effective "remote control" of the production during design and management.
(4) The design of separate, neutral functions by dedicated specialists, preferably by quantitative analysis.
(5) This preferably mathematical description does not fully correspond to reality and eliminates ethical considerations (aspects) during design by abstraction. It also can become counterproductive, because it tends to eliminate human creativity and imagination (Ferguson, 1992) and to replace it with quantitative design tools.

This method does not imply that engineering has fully become an applied (technological) science. As Vincenti (1990) has observed,

> airplanes are not designed by science, but by art in spite of some pretence and humbug to the contrary. I do not mean to suggest for one moment that engineering can do without science, on the contrary, it stands on scientific foundations, but there is a big gap between scientific research and the engineering product which has to be bridged by the art of the engineer.

However, Vincenti refers to creativity that bridges the gap, and he does not recognize the scientific method as applied by modern engineering and pursued by the technological sciences. Therefore Vincenti only refers to "pure" sciences (physics, mathematics) and does not recognize the systematic engineering method at work, e.g., in his description of the development of the countersunk rivet technology in aerospace industry. This development is according to Vincenti "empirical" and not modern. However, the typical engineering method is applied in this case. Most of the engineering design practice is "empirical" in the way Vincenti is using this characterization, and not based on "pure" science. But the powerful design method as described above is always the mindset of the modern engineer.

Whereas Van Riessen adopted Dooyeweerd's cosmology to analyze modern technology, his successor Schuurman (1995) mainly followed one of Dooyeweerd's other two lines: the critique of Western culture. His critique is clearly based on the implications of Van Riessen's structural analysis of technology for Western culture in which this technology plays a dominant role. Schuurman found that the characteristics of the modern worldview are similar to the characteristics of modern technology as described by Van Riessen. Modern society is designed and constructed in the same way as modern artifacts: by theoretical, abstract assembly from neutralized parts. Society is therefore treated as an assembly of separate individuals; nature is controlled and preserved as a factory; the body is healed and manipulated as a set of organs, and life can be constructed from chemical substances. Also in the arts one may distinguish the technicistic motive as an important driver, e.g., in the work of the Dutch painter Mondriaan. The modern Western worldview is technicistic because it reduces the multi-modal complexity of reality to a technically qualified object. This reduction becomes evident in the technical way in which we solve most of our problems.

This Reformational background is meant to clarify my position and to open my philosophical toolbox. Note that the Reformational philosophy of technology started a philosophical analysis of technology as a critique against the black box method of classical philosophers of technology, while Schuurman mainly studied the cultural impact of technology and technicistic thinking on society and culture, very much in the line of classical thinkers such as Heidegger and Ellul. I indicated that the classical and empirical ways of thinking about technology are linked in a structural analysis of modern technology.

2. AN AEROSPACE CASE

I will now try to relate the philosophical analysis of Van Riessen and Schuurman to my own professional situation, in order to make clear that Zandvoort's juridical turn does not fully acknowledge the complex and sometimes chaotic internal structure of modern technological development which creates the ethical vacuum that engineering specialists face in their profession. And it is especially this ethical vacuum that is recognized and allegedly solved by Zandvoort in a juridical way.

My main task at present is being the university project leader for the technology development program aimed at a new material for the fuselage of the ultra high capacity aircraft A3XX of the European aircraft manufacturer Airbus Industry. This huge aircraft is the European answer to the monopoly of the American competitor Boeing in the market segment of the Boeing 747 jumbo jet.

The A3XX is actually much larger. Because of the severe competition with Boeing, the pressure on the project to reduce cost and minimize weight is tremendous, and consequently the introduction of new, still unproven technologies is necessary. The A3XX will fly after 2004 and will be able to transport 550 to 800 passengers. Therefore safety is a big concern.

One of the candidate materials for the skin of the huge top section of the passenger cabin is GLARE: a GLAss REinforced aluminum laminate developed at Delft University with inherent high fatigue resistance, damage tolerance and burn through resistance (Vlot, Vogelesang and de Vries, 1999). We therefore advertise the material as being inherently fail safe. The laminate consists of thin layers of aluminum, bonded together by adhesive layers in which glass fibers are embedded. De-bonding of the layers and other internal degradation may cause a safety hazard and durability is therefore crucial.

The development of GLARE began in the early 1980s at Delft University. The material is currently produced for the European market by Fokker Aerostructures in the Netherlands. This company also owns the patents for the European market since January 1999 by taking over Structural Laminates Industries (SLI) from AKZO.

The development program for technology readiness of the A3XX is sponsored by the Dutch ministry for economic affairs. Airbus is a consortium of European partners in Germany, France, Spain, and Britain. Because of the "astronomic" production costs for the A3XX, Airbus tries to attract other risk sharing partners such as the Dutch aerospace industry, which is trying to find new markets after the bankruptcy of Fokker. The work share for the design and manufacturing of the A3XX is therefore not only full of technical but also of political hopes and hurdles. The safety of the A3XX, which is regulated by the airworthiness authorities, has new aspects because of the double deck layout and the new technologies that are applied. The future customers, the airlines, also participate in the design process and focus mainly on maintenance costs.

The tasks of Delft University in co-operation with SLI in this huge design process is the technology readiness of the new material by developing design concepts, computational methods, manufacturing techniques and quality control systems for the new material and structure for the top of the fuselage. The university researchers are the inventors and form the center of the knowledge of the new material. In the Dutch scene we have to co-operate with Fokker to transfer a part of our know-how, and with another aerospace laboratory in the Netherlands (NLR) which is partly a competitor of our university lab. Within Airbus co-operation is necessary in the framework of the delicate relationship of the partners, especially between Germany and France. The German company DASA and the French company Aerospatiale are both responsible for a section of the fuselage, but each company has different interests, cultures, and design methods.

Because of the severe competition between Airbus and Boeing, the company is extremely cautious with information. Delft University signed a secrecy agreement with Airbus via SLI/Fokker, and the knowledge that is gathered in the project is owned by the Dutch government.

My work is therefore at the moment not purely academic but is rather applied and oriented to the development of new technologies and methods. This is also due to the application-oriented approach of our university lab, which is rather unique and was copied from the situation in academic hospitals. In this way I can work as an academic researcher in an industrial environment, but I try to keep my freedom and objectivity. In this specific project I encounter the following ethical questions:

1. The A3XX is inevitable because of the unrestricted, explosive growth of air transportation that clearly conflicts with sustainability (in terms of noise and emissions). The application of GLARE will reduce fuel consumption compared to conventional aircraft structures, but it is likely that this reduction will lead to cost savings that will lead to further growth of air traffic (this opposite effect is called the "rebound effect"). I consider therefore my initiation of research into methods for recycling the hybrid, laminated material as the only contribution to sustainable development in the GLARE project.

2. The development and application of the material in combination with the design of the aircraft is situated in a huge and complex set of parties with different interests and cultures. GLARE is in a tough competition with other materials and each of the materials has advocates in different R&D and specialized engineering groups with other methods and cultures among the Airbus partners. The aircraft manufacturer is under extreme pressure in terms of time and personnel to meet the milestones in the development of the A3XX. The project is therefore far from smoothly organized. It is a chaotic and adaptive process in which communication is difficult and hierarchical power is the main driver. The advantages and disadvantages of the materials found by tests and calculations are translated into risks (in terms of financial and technical risks) and communicated within the organization. Hard data and figures become subjective estimates, gut feelings, and rumors. Carefully prepared demonstrations and presentations are important on different levels, from engineers to upper management, in order to promote the enthusiasm needed to create momentum

for the material within the organization. Unexpected drawbacks, disadvantages, and risks are interpreted as fixed or showstoppers by skeptics, and as problems that can be solved by believers. In this debate the past experience laid down in specifications and test methods play an important role. Because these are based on past experience, a fully new development like GLARE always needs reinterpretation and adaptation. This is a very difficult hurdle because of the inertia of the organization, and because specifications are theoretical and abstract tools in the handbooks of experienced specialists. To change them a new perspective is needed based on real life, i.e., on the actual behavior of the new technology.

Summarizing: In this chaotic co-operation of different partners and parties within the organization, communication is difficult, and objective material properties derived from test data and calculations are always expressed in a prejudiced context. Specifications, handbooks, and other design tools hamper changes. Our small university group on materials and structures is rather multi-disciplinary but has to deal with specialists within the industry with their own limited perspective: designers, production specialists, maintenance engineers, corrosion experts, etc., etc. And because GLARE is a mix of metal and re-inforced plastic, we have to deal with both related types of experts within those departments. The development of a new material in such a project has a major political aspect in which truth has a difficult life and the responsibility of an individual specialist is rather vague and limited.

3. Safety and environmental issues related to the uncertainties of a new technology are built up from interrelated disciplines, such as aerodynamics, avionics, manufacturing, structural design, material behavior, and maintenance, and are therefore rather complex. My expertise is limited to a small, specific part of the materials science and design of structures, and relations with other disciplines are hard to foresee, while safety and environmental issues are characteristics of the artifact as a whole. This is a special instance of the ethical "problem of the many hands". The supposition of Zandvoort that engineers in a company are "autonomous actors" who may be held responsible may be true within the boundaries of their abstract function and specialization, but whistle blowing and communication with the public in this complex environment requires crossing these functional boundaries of individual liability. Pinning engineers to their expertise, for which they may be held strictly liable as individuals, as Zandvoort proposes, can be counterproductive in an ethical sense.

I conclude that my ethical dilemmas stem from safety and sustainability issues in combination with problems of carrying responsibility as a functional specialist in a complex, chaotic organization with formal secrecy agreements. Will the new legal system proposed by Zandvoort stimulate me in responsible behavior?

3. JURIDICAL TURN?

There are three aspects addressed by the two claims of Zandvoort.

1. *Strict liability will stimulate sustainability and protection of the environment.* The growth of air transport is the driver behind the material development as described above. This is a cultural problem of our society based on economic growth, which leads to the fact that every saving will be spent for more growth (rebound effect). Liability will not and cannot affect the roots of this problem. Liability for environmental degradation can be useful for localized, traceable damage. This is an important difference with the example of the drugs industry used by Zandvoort, because the effects of a specific drug can be traced to specific companies. But who is liable for the greenhouse effect or holes in the ozone layer? In a way, the application of this new material will make structures environmentally friendlier but that is also because aircraft consume so much fuel during flight that any dirty, un-recyclable, energy-consuming material that saves weight will be beneficial for the life cycle of the structure as a whole. And more generally, a life-cycle analysis to determine environmental impact is a highly questionable, quantitative computer tool that can only give rough indications (Tempelman, 1999; Bras, 1999). The uncertainties in determining environmental impact, the shared responsibilities between specialists, and the rebound effect are bigger obstacles than liability. Moreover, liability can certainly be no answer for the destruction of nature in cases where no people are affected. E.g., aesthetical degradation and erosion of bio-diversity cannot be attacked because nature is not a partner in liability. How can they be valued and quantified?

2. *Safety will be promoted by strict liability.* Aircraft manufacturers are already strictly liable for design and production errors. However, because an accident is usually not only caused by a design or production flaw but by a failure of a complex chain of imperfect design, pilot error, inspection mistakes, maintenance errors, inadequate regulation or supervision by government agencies, etc., it is always hard to isolate causes and liable actors. However, the high safety standards reached in aviation are not established by this liability but by regulations and by ensuring an adequate safety organization before accidents occur and not by liability afterward. Liability is not the key-issue here – it can even lead to lower safety standards, if the price of an accident is lower compared to the costs to prevent it! What is vital is the triangle of manufacturer, airlines, and government agencies, operating under the pressure of public opinion. Within the drawbacks of the modern design method with a high degree of specialization, this triangle is able to keep safety high on the agenda and can enforce adequate regulations and measures. Safety is a crucial boundary condition for all aerospace engineers because of the organizational setting, but

also because it is such an important issue during education that it becomes a part of their mindset. Sustainable development and multi-modal ethics should also be internalized in organizations and people in the same way.

3. *For the right to be informed about risks a legal protection of the employee is necessary.* This may be in conflict with the secrecy agreements that are in place within the development of an aircraft material, as described above. However, this secrecy is absolutely essential. It is the corner stone for the development of a new technology, not only because of severe competition, but also because negative rumors will kill the development. I think that it can be supported by empirical evidence that the initial development of new technical ideas always proceeds in a shielded environment. The whistle blower of the ethical text books, who stands up and speaks, is an ideal. I am afraid that the foundation of this ideal in a juridical way will become a means for less courageous or less involved people to spread rumors and to kill new ideas before they get the chance to become mature enough to leave the development phase.

As discussed above, the information about complex technologies is never objective and is always vague and highly specialized. What is the meaning of informed consent in this respect? Due to the highly multi-disciplinary building-block approach of modern technology it is extremely difficult to introduce individual responsibility regarding safety and sustainability. Within the structure of modern technology, as analyzed by Van Riessen, there is always a truth expressed by the saying that "ethics is not my department", especially when you are liable. The figure of the whistle blower has to be cherished, but reasons to blow the whistle must be strong enough. It sure can require martyr-like qualities. The most difficult issue for the whistle blower is not the legal backing, but the certainty that there is sufficient reasonable evidence to blow the whistle. Even in cases where hindsight reveals sufficient evidence for whistle blowing, as in the Space Shuttle Challenger disaster, it was not legal protection nor strict liability but the technical uncertainties and the power games between specialists and managers within the complex project organization under time pressure that were the weakest links that lead to the catastrophe. ***Whether*** GLARE or the lack of GLARE in the A3XX will cause a disaster, it will probably be because of the same factors.

REFERENCES

Achterhuis, H. (1997). *Van stoommachine tot cyborg*. Amsterdam: Ambo.

Bras, R. M. (1999). *Adjusting Life Cycle Assessment methodology for use in public policy discourse*. Dissertation Delft University.

Dooyeweerd, H. (1969). *A New Critique of Theoretical Thought*. Philadelphia: PRPC.

Ferguson, E. S. (1992). *Engineering and the Mind's Eye*. Cambridge, MA: MIT Press.
Meijers, A. W. M. (1998). *Wat maakt een ingenieur?* [What produces an engineer?]. Inaugural lecture, Delft University of Technology.
Mitcham, C. (1994). *Thinking through technology*. Chicago: University of Chicago Press.
Noble, D. (1997). *The Religion of Technology – The divinity of man and the spirit of invention*. New York: Alfred A. Knopf.
Riessen, H. van. (1949). *Filosofie en techniek*. Kampen: Kok.
Schuurman, E. (1995). *Perspectives on Technology and Culture*. Sioux Center: Dordt College Press.
Snow, C. P. (1965). *Two Cultures: and a second look*. Cambridge University Press.
Tempelman, E. (1999). *Sustainable transport and advanced materials*. Dissertation Delft University.
Vincenti, W. G. (1990). *What Engineers Know and How They Know It – analytical studies from aeronautical history*. Baltimore and London: The John Hopkins University Press.
Vlot, A., Vogelesang, L. B., & de Vries, T. J. (1999). Towards the application of Fibre Metal Laminates in large aircraft. *Aircraft Engineering and Aerospace Technology, 71*(6), 558–570.

SELF DETERMINATION, STRICT LIABILITY, AND ETHICAL PROBLEMS IN ENGINEERING

Henk Zandvoort

1. INTRODUCTION

Same Problem Perception, Different Remedies?

In my chapter I stressed the necessity to adapt the existing legal systems to certain ethical principles, in order to diminish ethical problems of contemporary technological society and of contemporary engineers. For this reason, Vlot calls my contribution "a juridical turn". Vlot says he sympathizes with my intentions and ideas. But he thinks that my proposals "will not and cannot affect" the roots of major problems of present day technological society such as the non-sustainable character of many activities and the deterioration of ecosystems; that they do not "touch the complex and sometimes chaotic internal structure of modern technological development which creates the ethical vacuum that engineering specialists face in their profession"; and that they "remain external to some real engineering dilemmas within the structure of modern technological development". Finally, Vlot mentions the aircraft industry as a counterexample to my claim that strict liability promotes safety.

To illustrate his claims, Vlot presents a case from his own experience in aerospace research and development, and he describes the ethical problems of an engineer who is working in such a setting. Vlot's case-description is

The Empirical Turn in the Philosophy of Technology, Volume 20, pages 219–243.

ISBN: 0-7623-0755-2

exemplary for the organizational context in which many modern technological developments take place. His examples are relevant as well for activities taking place in other hierarchical organizations. Ethical problems such as those described by Vlot equally arise in, for instance, a governmental department. Whereas the subject matter differs (public administration and public policy rather than airplanes), the structure of the ethical problems is the same. My proposals are (in part) intended as answers to the question of how such ethical problems can be removed or diminished.

Vlot's and my own recognition of the ethical problems of professionals in (hierarchical) organizations appear to be similar. At the same time our conclusions concerning possible solutions seem to diverge. This inspired me to the response below. My aim is to show that, in spite of what Vlot apparently thinks, my proposals are extremely relevant to solving or diminishing the ethical problems of engineers relating to safety and sustainability.

2. RECAPITULATION OF MY PROPOSALS. FEASIBILITY AND RELEVANCE.

Self Determination, Strict Liability and the Right to be Informed on Risks Caused by Others

In my chapter I defended two related proposals, namely the introduction of:

(1) strict liability for technological (and other) activities; and of
(2) the right for everyone to be informed about risks caused by others, and derived from this, the duty of everyone who entertains activities that may create such risks, to inform all those who may potentially be struck by those risks.

The above order reflects that 1 requires 2, as will be explained shortly. At the basis of 1 and 2 is another principle, namely the principle of self determination, stating that all persons are free to do what they please as long as they do not harm others. (The right of each to be safeguarded against the consequences of the actions of others is equivalent to this.) I assume that all value judgements are subjective, and hence that harm is a subjective notion. The principle of self determination thus requires that all persons must be allowed to evaluate the harmfulness of the consequences that they experience, irrespective of who caused these consequences. A further implication of the principle of self determination is that actions are allowed only if those who may experience the (possible) consequences have given their consent. Strict liability, as defined in my contribution, is the duty to repair or compensate for harm due to actions that lack the informed consent of those who may experience the harm. It is the natural counterpart of self determination.[1] Informed consent requires that all have been fully informed on the human made risks that threaten them. Hence, if strict liability (principle 1) is to function properly, all persons should have the right to know all relevant information (principle 2). To realize this right requires that those who create risks should have the duty to inform those subjected to the risks.

I am aware of problems associated with this duty to inform. Which information is relevant? What is a risk? Which information should be given, and which not? The existence of such problems does not alter the necessity of a duty to inform. It merely shows that there is important work to be done in order to adequately implement it.

Feasibility and Relevance of My Proposals

I will distinguish two aspects:

(A) Relevance. How would my proposals, if implemented, solve or diminish the ethical problems identified by Vlot?
(B) Feasibility. To what extent is it at all possible to implement my proposals?

At some points Vlot's skepticism toward my proposals appears to concern their feasibility (B). At other points he denies their relevance for solving or reducing ethical problems of engineers, even if the principles would have been implemented (A).

I do not claim that it is easy or simple to fully or even partially implement my proposals. For that, the inherited situation is too far removed from a situation where my principles would hold. To properly understand the present state of the world, certain developments of the past two centuries should be taken into account. They took off roughly with the political (r)evolutions triggered by the Enlightenment, and were strongly influenced and enabled by the subsequent industrial development of the 19th and 20th centuries. I would like to mention four related aspects:

(i) the huge scale and scope of technological and industrial developments and their effects;
(ii) the enormously increasing role of (mostly hierarchical) organizations, both private and public, treated in the legal systems as (artificial) persons;
(iii) the creation of limited liability of these artificial persons and the introduction of various sorts of conditions to liability of (both artificial and natural) persons;
(iv) all this governed by a system of collective (political) decision making based on majority rule rather than on consensus.

These elements taken together have led to a situation very different from situations where my proposals would be fully implemented. However, much more is possible than what Vlot apparently assumes.[2] Here I do not mean possible in an abstract or utopian sense, but in actual social reality. I will show this in Section 3 below, where I give an overview of some important recent developments in tort law relating to modern technology. This overview serves as the basis for my response, in Section 4, to Vlots comments. I will end with a summary and conclusions.

3. RELEVANT DEVELOPMENTS IN MODERN TORT LAW[3]

Tort law is that part of private law dealing with wrongful acts for which damages can be obtained in a civil court by the person wronged, other than wrongs that are (or are considered as) breaches of contract.[4] Tort law is of special relevance to modern technological activities and their harms because here usually there is no contract between the actor and the one harmed, the breach of which led to the harm in question. Examples of such harms are: industrial and traffic accidents; nuisances due to traffic or industrial activities; health and environmental effects of toxic or otherwise dangerous substances; effects of CO_2 and other exhausts on climate and/or environment; effects of pollution in general; exhaustion of natural resources in general.[5] In agreement with this, the development of tort law in the last two centuries was closely related to the enormous development of technology in that period. Likewise, a look at present day tort law is very relevant for the issue of how technology may be managed in an ethically and socially responsible or acceptable way.

In recent environmental liability law, examples of relatively strict forms of liability can be found, as well as other relevant concepts such as market share liability, multicausality and multi-actorship, and class actions. Many of these legal concepts and theories were originally developed for cases of so called toxic torts in the sphere of human health, such as the cases of the carcinogenic medicine DES and of asbestos. Some judges, lawyers, and legal writers try to extend such concepts and theories to other domains such as environmental damage.[6]

Terminological Note

In my chapter, strict liability was defined as the principle that persons are unconditionally liable for (i.e., is unconditionally required to repair or compensate for) the possible consequences of their actions to others who had not given their informed consent to these actions. The principle does not entail something about liability with regard to persons who did give their informed consent for the activities under consideration.[7] In positive law, strict liability only refers to the unconditional (legal) duty to repair or compensate for damage due to certain unlawful activities, i.e., irrespective of whether *the particular persons damaged* had or had not given their informed consent to the activities. Wherever confusion might arise, I will indicate below how the term strict liability is meant.

3.1. Strict Liability for Environmental Damage

An example of a rather strict form of liability for technological activities is the Oil Pollution Act (OPA) in the USA. This law was enacted in 1990, in response to environmental accidents with oil vessels, especially the Exxon Valdez. The law applies to vessels or facilities from which oil is discharged. According to one author this law has "created the realistic prospect of absolute liability, and, perhaps, in many circumstances, no right to limit liability" (Force, 1993, p. 21). With absolute liability the author means that considerations of so called "fault" are irrelevant for the existence of liability. "No right to limit liability" means that the Act has no legal provisions that the actor may use to limit liability to certain pre-set bounds.

The OPA distinguishes three groups of persons who may be liable: responsible parties (including the owner and operator of a vessel or facility), guarantors, and third parties. A guarantor is any person other than the responsible party, who has provided evidence of financial responsibility for a responsible party (p. 29). A third party is a person other than the responsible party or its agents or employees, or the agents or employees of its independent contractors, or a guarantor whose conduct is the sole cause of a discharge of oil.

The OPA makes owners liable for a wide range of possible effects of oil discharged from a vessel or facility. It permits claims not only for the cost of clean-up and restoration of national resources (such as coastal bays). Also, a governmental entity may recover damages for injury to, destruction of, loss of, or loss of use of, natural resources, including the reasonable costs of assessing the damage (p. 32). And any person who uses natural resources which have been injured, destroyed, or lost may recover damages for loss of subsistence use of these natural resources, without regard to ownership or management of the resources (p. 33. See also p. 70).

The OPA furthermore requires responsible parties "to provide evidence of financial responsibility" (p. 41). Vessels for which this requirement has not been fulfilled are denied clearance or entry.[8] The amount of financial responsibility actually required should be sufficient to meet the maximum amount of liability to which the responsible party would be subjected in a case where it would be entitled to limit its liability under certain provisions of the Act.[9] The financial responsibility requirement may be satisfied "by evidence of insurance, surety bond, guarantee, letter of credit, qualification as a self-insurer, etc." (p. 42).

Any claim for removal costs or damages authorized under the Act may be brought *directly* against the guarantor who provides the evidence of financial responsibility for a responsible party (p. 42).

3.2. Assessment of Damage to Natural Resources

In order to provide liability for damages to natural resources with operational meaning, methods for assessing damages are needed. This topic is addressed in, e.g., De la Rue (1993). Following a pattern developed under the general environmental laws of the USA, notably the Comprehensive Environmental Response Compensation and Liability Act of 1980 (CERCLA), the OPA provides for so-called natural resource damage assessment (NRDA) regulations to be drawn up, to govern compensation awards for claims relating to environmental loss or damage (p. 70).

An interesting early case of environmental damage assessment was the *Zoe Coloctroni* case, which was heard in 1980 by the first circuit of the U.S. Court of Appeals (p. 71). The matter concerned an oil tanker that had run aground as a result of her unseaworthy condition, causing pollution on the coast of Puerto Rico. At first instance, the government authorities of Puerto Rico were awarded some U.S. $6 million. Of this only $78,000 was needed to compensate for the costs of cleaning up the oil spill. Five hundred thousand dollars was awarded as the cost of replanting mangroves, and some $5.5 million as the cost of replacing marine organisms killed by the spill. The judge arrived at the latter number as follows. He observed that marine organisms had been killed in an area of about 20 acres. Surveys conducted by the plaintiffs convinced him that there had been a decline of about 4.6 million organisms per acre, which meant that 92 million marine animals had been killed. The judge estimated that the cost of replacement of these animals, which were commercially available, would amount to minimally $5.5 million (p. 72).

In the *Zoe Coloctroni* case, the Court of Appeals refused to endorse this approach, saying that it did not express the right sense of proportion. It accepted as recoverable costs only those "reasonably to be incurred by the sovereign or its designated agency to restore or rehabilitate the environment in the affected area to its pre-existing condition, or as close thereto as is possible *without grossly disproportionate expenditures*" (p. 72).[10] However, the NRDA regulations were introduced only after this case, and the author of the chapter expresses the view that "On the face of it, these (i.e., the NRDA regulations) seem to call for a return to the more rigid methods of calculation favored at first instance and rejected by the Court of Appeals" (p. 72). He adds to this that an explanation for the Court of Appeal's judgement that the original calculation lacked the proper sense of proportion may be that in the *Zoe Coloctroni* case the Puerto Rican government agencies involved had no intention to actually purchase the 92 million marine organisms. But some legal theorists argue against this, saying that claims should be accepted for the estimated cost of reinstating the environment, irrespective of whether the supposed measures of reinstatement are actually carried out (p. 75).

Another interesting method for assessing damage was proposed by the U.S. National Oceanic and Atmospheric Administration (NOAA). The procedure comes down to this. Ask any individual who can be considered to have an interest in this piece of nature lost, what it would have been worth to him to retain it. The sum of these valuations is taken as the total value lost (De la Rue, 1993, pp. 75–76). Writing in 1993, De la Rue expected that the procedure would be included in detailed regulations applying NRDA claims under the Oil Pollution Act.

The approach described has been called "contingent valuation methodology (CVM)". If implemented, it might work as follows. Suppose an average USA family is found to be prepared to pay $ 100 in order to preserve the natural beauty of Prince William Sound in Alaska (for instance, to retain its availability for themselves or for their offspring). Because there are some 100 million families in the US, the conclusion might be that compensation of $ 10 billion should be paid for the spill from the Exxon Valdez (p. 76). De la Rue doubts however whether it is really desirable to introduce this kind of methodology to assess compensation claims under civil tort law, because, so he says, such claims are more akin to criminal fines, ending up as a windfall to the public treasury (p. 76). This argument would not hold if the damages would be paid to the citizens directly. Neither would it hold if the damages would be earmarked for restoration or creation of natural resources, and if their destination would be determined using the same type of methodology that was used in the first place to assess the damage.[11]

3.3. Multiple Causation

If several actors contribute to the same damage or loss, this is called multiple causation. Although there are old examples of multiple causation, the phenomenon has multiplied under the influence of technological development in the 19th and 20th centuries. Multiple causation is typical for much damage due to technological activities, e.g., a river polluted from several sources upstream or a forest damaged by acid rain.

An important question in cases of multiple causation is whether the liability should be several or joint. In the case of several liability, different actors are liable for their individual contributions to the loss, but not for more. In the case of joint liability, the complete loss is recoverable from any actor (whose concern it then becomes to seek redress with co-actors). In a number of Western European cases of technological torts, including environmental damage, the courts have adopted joint liability rather than several liability (Van Dunné, 1993, pp. 376–377, "Rechtsvergelijking"). This was based upon the argument that it would be unjust to load the victim, not having been involved in the activities leading to their loss, with the task to localize all the actors in order to recover damages.

3.4. Market Share and Risk Share Liability

The concept of market share liability was first developed in the context of so called "toxic torts". Legal scientists use "toxics" as a catch-all term that may include asbestos, pesticides, herbicides, insecticides, drugs, chemicals in general. It even has been extended to radiation and micro-organisms.[12] Often, such toxics cannot be traced back to a specific manufacturer. Market share liability was developed to deal with this problem.

Market share liability was first articulated by U. S. A. courts, in connection with the medicine DES: "Faced with a class of innocent DES-daughters, an industry where over 200 manufacturers of DES had existed, and five defendants who collectively had made 90% of the product, the court held that as long as the plaintiff had sued a sufficient number of manufacturers to constitute a substantial share of the market, then each defendant could be held responsible for a proportion of the judgment based upon its market share unless it could exculpate itself" (Harrigan, 1993, p. 48). In most of the USA-cases, the liability was several, not joint: each defendant was liable only to the extent of its market share (*ibidem*).

Some U.S. states have deviated from the above in interesting ways. For instance, Wisconsin adopted in 1984 a theory whereby each manufacturer who contributed to a *risk* of injury can be held responsible. A plaintiff has to sue only one defendant. It then becomes the defendant's responsibility to identify and bring into the case other possible responsible parties or bear the exposure alone. The idea behind this theory is sometimes called *risk share liability (ibidem)*. With respect to DES, the Dutch High Court similarly decided in 1992 that each of the DES-manufacturers summoned in court was jointly liable for the whole of the loss (Van Dunné, 1993, pp. 388–389; Kröner, 1993, preface). Still different is a theory adopted in New York in a DES case, which is described by some authors as a combination of risk-share and market share liability. In this case, liability was determined by the "overall culpability" of each manufacturer measured by the risk it created to the public at large; each defendant's proportional share was based on its percentage of the national market (and limited to that, since there is only several liability). But there was no exculpation: even if a defendant proved it had nothing to do with the product responsible for the injury, the defendant was required to pay, in proportion to his market share (Harrigan, 1993, p. 49).

3.5. Generalization of Market Share and Risk Share Liability to Other Technological Torts

The possibilities of extending these concepts of market share and risk share liability to environmental torts are intensively being explored by legal theorists and lawyers; see e.g. Van Dunné (1993, p. 384 ff.) and Snijder (1990). Results can be seen in, e.g., the way federal and state regulatory agencies in the USA deal with the clean-up (remediation) of hazardous waste sites. Thus, CERCLA

> allows the government to clean up the site and charge the cost to the owners, transporters, and generators of the hazardous substances. As a practical matter, the government tries to identify all "potentially responsible parties" ("PRP's") and get them to clean up the site themselves [. . .] PRP's prefer it this way because anything the government does tends to cost double or triple what the private sector could achieve. A PRP who is liable under CERCLA is strictly liable, it is no defence that the company was not negligent or that it complied with all applicable state and Federal laws. The amount of material shipped or found is immaterial for liability purposes. Joint and several liability applies. Nor is it material that the hazardous substance shipped to the site by a PRP is not being (or likely to be) released. All parties who disposed of any hazardous substance are liable (Harrigan, 1993, p. 49).

This approach strongly resembles the Collins Risk Share theory, where merely contributing to the risk of harm can render one liable for all the harm.

Since the government and most PRPs are more interested in settling than in litigating, most responsible parties settle for a "volumetric" share, that is, they contribute to the cost of remediation a percentage based upon the ratio of the volume of the substance they shipped to the total volume of all hazardous substances on the site. This is analogous "market share" liability, where "volumetric" is substituted for "market" (Harrigan, 1993, p. 50).

3.6. Class Action

Not only the number of actors, but also the number of victims may be large in typical cases of damage due to modern technological activity. Often, as in the case of environmental pollution, a large number of persons is confronted with a damage that for each individual is relatively small as compared to, e.g., the costs or efforts connected with a trial. This means that an evaluation in terms of the expected costs and the expected benefits of a trial usually leads for each individual to the decision not to sue. So there is no redress, although the total damage may be large given the large number of victims. Indeed, the total harm or disadvantage of those affected may exceed the total advantage of the polluter(s) from not taking precautions to prevent the pollution.[13]

In order to make (private) tort law relevant to such cases, the possibility of so called class action is being introduced in some cases of environmental legislation (for instance in the EC Draft Directive on Liability for Damage caused by Waste). The essence of a class action is that (a representative of) a group is given the right of legal action for the members of the group jointly. According to Sellschopp (1993, p. 54–55), class action or "popular action" was part of Roman law, and is still found in the civil law books of Chile and Colombia.

3.7. Burden of Proof

A widely accepted condition for liability is that a causal relationship should exist between an act of a defendant (or his agents) and the damage from which a victim wants redress. This condition agrees with the time honored ethical principle that it is unjust to blame persons for something they had no part in. However, for technological activities it is often difficult to establish the causal relationship between a damage or loss, and the act(s) that caused it. Examples were discussed above in connection to DES and environmental pollution. These problems of proof are caused by at least the following circumstances. As a rule, (1) several actors contribute to one effect, (2) effects have a probabilistic nature, and/or (3) available knowledge about causes and effects, be it causal/nomological or statistical, is partial or lacking.

Because of this, the outcome of a trial and hence the factual nature of the liability imposed, may depend crucially on who carries the burden of proof: whether victims must prove that defendants caused damages, or whether defendants must prove that they did not cause damages. Many developments in tort law display a (partial) shift of this burden of proof from the victim to the defendant. Examples were given above in connection with the medicine DES. For other developments relating to burden of proof see the discussion of strict liability in Zweigert and Kötz (1987, ch. 49). As is shown by these authors, 20th century judges and legislators have in several cases decided to shift the burden of proof from the plaintiff to the defendant, especially for activities considered risky, or more risky than usual.

The issue of burden of proof is (together with the issue of "admittance to court") a fundamental subject. It is closely related to the issue of scientific and technological uncertainty and ignorance, especially in combination with the phenomenon of continuous technological innovation. It will also be clear that a duty to inform, as was defended earlier, has an important bearing on the subject.

3.8. Non-Additive Effects

I end this overview with an issue that, as far as I know, is not discussed as a separate issue by legal theorists. I address it nevertheless because it may look like a serious problem.

In the case of liability for environmental pollution, the typical situation is that of non-linearity of dose-effect relations. One may ask whether this does not lead to new juridical problems. For suppose P would be the only polluter, and that P's pollution of quantity A would yield an environmental damage xA for which P is liable. Suppose now that P is not the only one, but that Q is producing a pollution B. Then, because of non-linear effects, as a rule the total damage will exceed x(A + B), say by a factor y > 1. In order to cover the damage, P now should be liable for yxA which is more than the original amount xA. Is this reasonable?

On closer inspection this case is not different in structure from, for instance, the case of two drivers concurrently causing an accident that strikes a third party. The non-linearity of the dose-effect relation in the latter case is even extreme. For if only one of the drivers would have been negligent, there might not have been an accident at all. Nevertheless, people usually do not question the fairness of holding the drivers liable for the actual damage, rather than the damage they *would have* caused if the other would not have been there.[14]

4. RESPONSE TO VLOT'S COMMENTS

Strict liability and ethical problems of engineers

If the principle of self-determination is accepted, the ethical problems of engineers can be solved completely (and reduced to "ordinary" problems of decision making) only if their activities would have the informed consent of all who may possibly be affected. At present, for many activities this consent has not actually been given. It is an open question for which activities, and under which restrictions and conditions, this consent is actually obtainable. Such restrictions and conditions might include requirements of safety and sustainability, and liability regulations for damage that might nevertheless occur.

If the principle of self-determination is accepted, strict liability (in the sense of being required to repair or compensate for damages) should pertain at least to activities not having the informed consent of all possibly affected parties. But this in itself would not solve all the ethical problems of engineers, because many possible consequences of actual engineering activities cannot be (completely) repaired or compensated. Think of deaths and irreversible environmental changes.[15]

However, the introduction of strict liability (in this sense) would certainly help enormously. This is not merely because victims would then at least be compensated for damage resulting from activities not having their consent. Also, and related to this, these activities would probably become more acceptable to those who are exposed to the possible consequences. Furthermore, as was remarked in my paper, the interests of the organizations that employ engineers would become more in agreement with ethical codes of conduct for engineers.

Having made these general remarks on the bearing of strict liability to ethical problems of engineers, I will now turn to Vlot's specific objections and questions.

4.1. Liability, Sustainability, and the Environment

Vlot says that liability will not and cannot affect the roots of the problems of present day society with respect to sustainability and the environment. By way of illustration, he describes an ethical problem encountered by an engineer. This engineer devotes his energy and talents to the development of a new material for constructing airplanes. If successful, his work leads to more efficient, because lighter, airplanes. The cost savings generated from this will stimulate further growth in air traffic, further undermining sustainability. The engineer thus contributes to a development which is considered (both in general and by the engineer, I suppose) as bad, or undesirable. Vlot thinks, apparently, that my proposals concerning strict liability are irrelevant to this ethical problem. I will try to show the opposite.

Strict Liability and Sustainability

In Vlot's view, non-sustainable phenomena such as the growth of air transport are essentially a cultural problem of contemporary society based on economical growth. Vlot wonders what difference strict liability would make for sustainability, saying that "liability will not and cannot affect the roots of this [cultural] problem." I could not disagree more. If the producers and/or users of airplanes would be strictly liable for all risks and nuisances that they create for others, this would have huge effects on those activities, and hence on the activities of engineers such as the ones described by Vlot.

Risks created by airplanes (either on their own or jointly with other activities) include:

- the risk of accidents such as the 1992 crash of an El Al Boeing on the outskirts of Amsterdam;

- the risk of climatic change due to CO_2 exhaust;
- the (known and unknown) effects of exhaust of substances such as NO_x, SO_2, etc. both at high altitude and in the vicinity of airports.

Involuntary nuisances include noise from airplanes and its various consequences, including but not limited to diminishing economic property values.

The actual legal situation is, of course, far removed from strict liability. To start with, many of the relevant laws (concerning permits of airplanes to fly over urban areas, concerning safety regulations, concerning the availability of relevant information such as the nature of a cargo being carried, and concerning liability in case of accidents) obviously do not have the consent of all concerned. Worse, the tax laws even force people who disagree with those activities to contribute financially. Nevertheless, there is no liability for the production of CO_2 and other exhausts; at best partial liability for accidents; and virtually no liability for other nuisances or damages due to air traffic.

Introducing stricter liability for these various consequences would no doubt lead to increasing prices of airplane tickets. Other things being equal, intensity of air traffic might diminish, and alternative technologies might be supported, such as information and communication technology, or alternative forms of transportation. (Possibly, Vlot's new material may prove useful for building Zeppelins as well.) But also if air traffic would nevertheless keep growing, the ethical problems of the engineer depicted by Vlot would be reduced, for the reasons mentioned above.

Example: Market Share Liability for CO_2 Production

The introduction of market share liability for the effects of CO_2 production seems quite feasible from the perspective of legal practice and thinking. The legal concepts required are available. They have been put to test in real laws, and are merely waiting to be used in broader contexts, so to speak. Seen from this perspective, it is feasible to discuss at the next U.N. conference on climatic change a proposal to introduce market share responsibility for CO_2. Such a proposal might help to convince developing countries of their sense of responsibility, and might increase their willingness to consider possibilities to reduce the CO_2 exhaust of their own developing economies.[16]

To add to the credibility of such proposals, a global fund might be created (again on the basis of market share, of course) that countries could appeal to if they would be actually struck by consequences of climatic change, e.g., for constructing dams, etc. This would be analogous to the "evidence of financial responsibility" of the OPA (see section 3.1 above).

Liability for Environmental Degradation; Vague Causes and Vague Effects
According to Vlot, liability for degradation of the environment can [at most] be useful for localized, traceable damages. I interpret him as saying that the introduction of liability for environmental damage is only possible or makes sense if damages can be localized and traced back to their sources.

Now it is true of modern technological activities that often both the effects (damages) and their causes are more or less vague or probabilistic, spreading widely over space and time, and/or are cumulative in nature. Traditional ethics, stemming as it does from pre-industrial times, is not accustomed to this. It is a matter of fact that traditional ethics, and its traditional implementation in law, function poorly with respect to contemporary technological activities.

As some have pointed out, traditional ethics and law work well for regulating the relationships between neighbors, but not for many technologically enabled relationships (via, e.g., CO_2 production) with the rest of the world. This is one of the ways in which ethics and law are lagging behind scientific and technological development. This is a very serious flaw, because in the summation these "long distance", technology-enabled influences of people upon each other may far exceed the influences of neighbors upon each other.

Nevertheless the developments sketched in Section 3 above show that numerous legal concepts relevant to these vague, widely spread, and probabilistic causes and effects of contemporary technological activities are actually available or under development.[17] There is nothing that precludes the further development of these or still other concepts. So I disagree with Vlot's conclusion that the phenomena of vague and multiple causes and of vague effects stand in the way of the effective introduction of strict(er) forms of liability for (technological) activities.

Implications of Uncertainties in Determining Environmental Impact (or Other Possible Impacts)
Vlot says the uncertainties in determining environmental impact are a bigger obstacle than liability. I see two aspects of what Vlot may have in mind here. The first aspect is that it may be difficult to assess damage done. The exposition in Section 3.2 above shows that there are ways to overcome or diminish such difficulties. The second aspect is that the *possible* damage from certain activities may be (and often actually is) far bigger than what the actors can bear. To the extent that damage can be repaired at all, it would suffice to demand adequate "evidence of financial responsibility" (see Section 3 above). By contrast, if the possible damages or losses are irreversible, the only way to avoid unbearable responsibilities is to obtain the informed consent of all who may possibly be harmed, *before engaging in those activities*. Vlot also mentions the "shared

responsibilities between specialists" as an "obstacle bigger than liability". I will deal with shared responsibility in Section 4.3 below.

Valuation of Natural Beauty (and of Other Natural Resources)
Vlot remarks that liability cannot help to prevent destruction of nature in cases where no people are affected, such as in the case of aesthetic degradation and erosion of bio-diversity. According to Vlot, nature is [according to the ethical principles invoked by me] no party to be respected. Here I also disagree, since natural values can be (and, by the way, can only be) appraised, or expressed, or safeguarded, by humans. For examples of possible procedures to assess the value of natural resources I refer to Section 3.2 above.

4.2. Liability, Safety, and How Safe Is Safe Enough

Vlot says that already now aircraft manufacturers are strictly liable for design and production errors and that moreover the high safety standards in aviation are not established by this liability but rather by regulations and by ensuring an adequate safety organization before accidents occur.

As is apparent from the above exposition, I do not deny the relevance or necessity of safety regulations and procedures. I merely claim that in addition liability should be strict, especially when there is no (informed) consent of those who may be struck. The basic ethical question here is whether it can be justified at all to impose a risk upon someone without his or her informed consent. To quote a report written more than two decades ago: "Under what conditions, if any, is someone in society entitled to impose a risk on someone else on behalf of a supposed benefit to yet others?" (Council for Science and Society, 1977, p. 37). I do not know of any answer to this question, other than the condition of informed consent. When that condition has not been fulfilled, liability should be strict, that is, repair or compensation should be as complete as possible. In cases of accidents such as the El-Al cargo airplane that crashed on the outskirts of Amsterdam, a viable approach would be to hold the "possible responsible parties" or PRPs (see Section 3.4 above) severally and jointly liable. In this particular case the PRPs might include the aircraft manufacturer, the aerospace company, and the relevant government agencies.

Vlot says that strict liability might even lead to a decrease of safety, as when the price of accidents (including liability costs) would fall below the costs that must be made to prevent the accidents. If those who are subjected to risks have given their informed consent such an outcome need not be objectionable. Ultimately, the acceptability of risks can only be determined by those who are subjected to those risks.

Vlot also says that safety "is a crucial boundary condition for all aerospace engineers because of the organizational setting, but also because it is such an important issue during education that it becomes a part of their mindset". I am willing to believe that the training of engineers imbues them with a special regard for safety. But this does not undermine the following: (1) In spite of the facts Vlot points to, safety does not function as an absolute value in the professional work of engineers. A sufficient reason for this is that absolute safety in (almost all) engineering activities is practically impossible. In practice, trade-offs between safety (of others) and other values or interests are constantly being made. The fact (also mentioned by Vlot) that it may often be difficult or impossible to make objective estimates of risks does not alter this. (2) As Vlot himself so vividly describes, engineers have only a limited influence on the processes of technological development in which they are taking part, and virtually no influence on the ways the outcome will be used. The processes of technological development in which they are involved are, moreover, characterized by chaotic elements, difficult communication, and the use of hierarchical power. It would be miraculous if in such a setting the mere devotedness of engineers to the value of safety would be sufficient to keep technology within safe bounds. To give an example from the parliamentary investigation into the El-Al airplane accident: the presumed fact that El Al maintenance technicians were devoted to safety did not prevent them from being occasionally overruled in safety matters by their superiors.[18]

For the same reasons I do not think that the "internalization" of sustainability and "multi-modal ethics" (whatever precisely Vlot may mean with the latter term) in engineers would be sufficient to solve sustainability and other ethical problems of technology. Apart from this: How can safety and sustainability be internalized in organizations, since they are not humans? This leads me to Vlot's third group of remarks and questions.

4.3. Responsibility In and Of Organizations

In the organizational context characterized by Vlot, the actual responsibility of an individual specialist is both limited and vague. Many important decisions are taken at levels up the hierarchy, and it is unclear for which aspects engineers are responsible, and what that responsibility comes down to. This is the problem I want to address below.

Hierarchical Organizations and Limited Liability

As was said in the introduction, this situation is not at all unique to engineers. It occurs in all contemporary hierarchical organizations, technological and non-technological, private and public. In view of the type of the ethical problems discussed by Vlot and me, I will only address here organizations that through their activities may harm persons outside. (Hence, I will not discuss organizations such as the association of amateur gardeners, the whist-club, and the Masonic lodge.) At issue is how responsibility for the effects of such organizations upon persons outside should be divided among persons inside an organization, that is: Who should be responsible for what? The core problem is that many contribute to a single outcome (an airplane disaster, the deterioration of (part of) the environment – while it is not so clear who contributed what. This situation was baptized "the problem of many hands" by Dennis Thompson (1980) (see also Bovens, 1998, ch. 4).

With few or no exceptions, both government and business organizations are characterized by a hierarchical structure of command and control. This structure is the legally embedded answer to the problem of many hands inherited from history. Woodrow Wilson, writing in the U.S.A. in 1887, believed that for governmental organizations, the solution was to be found in hierarchical responsibility. He argued that

> There is no danger in power if only it be not irresponsible. If it be divided, dealt out in shares to many, it is obscured; and if obscured it is made irresponsible. But if it be centered in the heads of the service and in the heads of branches of the service, it is easily watched and brought to book (Wilson, 1925 [1887], pp. 148–149).

So the idea was to make the highest in the hierarchy responsible, and to require strict loyalty of all others in the organization to this one person responsible for their actions. Similar ideas were put forward in the U.S.A. by Frank Goodnow and in Europe by Max Weber.[19]

The hierarchical mode of dividing responsibility in an organization has been implemented for government organizations, being the type of organizations with which authors such as Wilson and Weber were concerned. But this has become the dominant form of private commercial organizations as well, and is related to the juridical construct of legal persons, being the entities which "by default", that is, in the absence of grossly negligent or reckless behavior of persons in the organization, are held legally liable for the acts of the (persons in the) organization.

As subsequent experience has shown, this solution to the problem of many hands in organizations is far from flawless. In the following, I will restrict myself to private corporations.[20] One of the flaws is that their liability has been limited.[21] This, together with the conditional (as opposed to strict) liability for the activities,[22] can explain very much about how and why society arrived at its present situation.

Non-Hierarchical Organizations and Shares of Responsibility

I want to say a few things about possible remedies. For the time being, I will assume that corporations can be adequately treated as (fictitious) persons in the law. If the principle of autonomy is to hold for (at least) persons of flesh and blood, then corporations should regain unlimited liability, and should be subjected to the general principles and consequences of strict liability described above. This means that for all activities of corporations informed consent must have been obtained for all who may possibly be hurt. As was explained above, this consent, if obtainable at all, may well be subjected to provisions and conditions such as safety and liability regulations, possibly including requirements of "evidence of financial responsibility" (see Section 3 above).

This would hold for all corporations, irrespective of their internal structure. But hierarchical organizations might have more difficulties in obtaining the required consent than non-hierarchical organizations operating on the basis of individual responsibility and freedom. Hence, under the conditions specified, present hierarchical organizations might disappear, to be replaced by or transformed into non-hierarchical organizations.

Such non-hierarchical organizations might operate on the basis of "shares of responsibility" for their activities, in the following way. Suppose an organization is allowed to perform a certain activity under the condition of the availability of an amount X of evidence of financial responsibility. This amount may be divided in shares, allotted on the basis of free will to persons inside the organization (the present employees and directors) and outside (e.g., the present shareholders or guarantors). If the employees (such as engineers) are sufficiently free, they will accept such shares only if they think that the conditions and competency allotted to their task are adequate for performing the task, and if they are sufficiently compensated (in the form of wages and/or shares in profits) for the work done and for the personal share in the risk of the operations. The willingness of persons outside the organization (hence having no direct control) to take shares will increase with the willingness of insiders to participate.[23] In this way, it may be easier for a non-hierarchical organization A to provide a certain amount X than it is for a hierarchical organization B. In addition, the public might in the case of A be satisfied with less strict safety regulations, since the public might have more faith in A, for the same reason that external shareholders might have more confidence in A.

It must be clear that it is not simple to introduce individual responsibility for the safety and sustainability of the operations of contemporary organizations. Thus, the introduction of the "shares of responsibility" described above would require drastic changes in the ways these organizations presently function. This in turn would require fundamental changes at several places in the legal systems that, as a matter of fact, determine the ways organizations function. These changes would at least concern liability law and, relating to this, the procedures of collective decision making concerning, e.g., safety and liability issues; corporate law in particular and the law of legal persons in general; labor contract law; and laws regulating secrecy, freedom of speech, and the like. All these changes are extremely relevant to if not required for the solution of the ethical problems discussed by Vlot. So clearly feasibility is an issue here. But it should be noted that a series of small steps might lead to a large change, provided that the steps are in the same direction.

To illustrate some of the things pointed out above, consider the Space Shuttle accident and the decisions that led to it (see, e.g., Vaughan, 1996). According to Vlot it was the technical uncertainties and the power games between specialists and managers that led to the accident, not issues of legal protection of whistle blowers or liability issues. This is misleading, because such legal issues have a profound influence upon the way technical uncertainties are dealt with, and upon "the power games between specialists and managers". At the teleconference between the functionaries of NASA and the contractor Morton Thiokol that preceded the launch, the contractor management had essentially two options. It could either recommend to launch, or recommend not to launch, the latter amounting to a veto against the launch. According to rational decision theory and given their legal position as agents of Morton Thiokol, the task of management was to assess the possible consequences (including chances) of both options, determine the (positive or negative) value of both options *for Morton Thiokol*, and choose the option with the highest valuation. Looking at available accounts of the teleconference, it is plausible that this is roughly what management did.

Hence, the outcome depended, among other things, upon the liability consequences for Morton Thiokol in case of a crash. (In reality, there was a contractual fine in dollars in case of a crash.) The outcome might also have been different if the managers and or the engineers would have held "shares of responsibility" of the type proposed above.

It will be clear that, if both the engineers and the managers held shares of responsibility for the same activity (the launch of the space shuttle) the only rule of internal decision making acceptable to all would probably be consensus, or, what amounts to the same, right of veto for all.[24] This may be compared with the power of organizations, rather than the power of individuals making up an organization. Among organizations a right of veto is often actually the case. In the Space Shuttle case, for instance, the contractors of particular parts of the Space Shuttle, of which Morton Thiokol was just one, all had a veto right on the launching decision.

More can and should be said about the possible internal structure of non-hierarchical organizations with external effects. But at this place I cannot pursue the issue any further.

4.4. Duty to Inform Versus Secrecy

Vlot says that in technological developments as described by him,

> secrecy is absolutely essential. It is the cornerstone of the development of a new technology, not only because of severe competition, but also because negative rumors will kill the development.

In response I want to point out that the principle of self determination only requires (and justifies) the right to be informed for activities that may possibly harm others. In principle this leaves room for secrecy in the pre-market stage of a product. But this cannot justify all forms of secrecy in the pre-market stage. Think of the development of genetically modified organisms, and the environmental and other risks these may embody once they have been engineered (they may escape from the laboratory, be stolen by criminals, etc.). If there are such risks already in the stage of research and development, my duty to inform requires that these be made known to those who are subjected to the risks.

As was already recognized, it is not simple to determine what forms a possible risk for others and what not, and hence what should be communicated and what not. But this is clear: without communication, there can be no informed consent. The principles put forward above imply that strict liability should in any case apply to risks that were considered non-substantial and for that reason were not communicated, but that nonetheless materialize.

However, as was pointed out earlier, this strict liability alone will not completely solve ethical problems relating to risks, because, if technological risks materialize, the effects are usually irreversible (e.g., deaths). Hence, more openness and transparency of organizations having such serious impact on the world outside is essential if we want them to function better than they do right now, and if we want to prevent still greater problems in the future. This can only be brought about by giving such organizations and their employees a duty to inform as was discussed earlier.[25]

Whistle blowing is in one respect merely a symptom (but a serious symptom) of fundamental problems with present day organizations. The requirements imposed on organizations and on the people working in those organizations should be such that whistle blowing either is not necessary, or else takes the form of an entirely normal and legitimate aspect of the job.

Consider once more the ethical codes of engineers. These codes suggest that the duty to inform the public about possible risks etc. is a normal duty for an engineer. This is rightly so in view of the principles and analyses put forward above. At the same time, living up to such a duty in actual reality will often require abnormal actions that, if not plainly illegitimate, are almost always extremely disadvantageous to the actor. Similar things hold, as was stressed before, for many others working in hierarchical organizations, both private and public. Given the overwhelming role played by these organizations in contemporary society, the persisting existence of such fundamental inconsistencies may be extremely harmful and dangerous to society at large.

5. CONCLUSIONS

Vlot and I apparently have a shared perception of ethical problems in engineering relating to safety and sustainability issues. But we disagree about the feasibility and effectiveness of my proposals to solve or reduce such ethical problems (Section 1). These proposals come down to the introduction of strict(er) liability for technological activities, and to the introduction of a duty to inform others on possible risks caused by activities (Section 2). I have tried to answer Vlot's criticisms and comments, and have tried to show by examples and elucidation that my proposals are both feasible and relevant. To underscore their feasibility, an overview of developments in recent risk and environmental liability law was provided, showing that already now relatively strict forms of liability and related concepts are being practiced at some places in actual positive law (Section 3). Next, I have argued for the relevance of my proposals for dealing with ethical problems of engineering and technology, relating to respectively sustainability and the environment (Section 4.1); safety (4.2); responsibility of and in organizations (4.3); and the right to be informed on risks caused by humans (4.4).

I conclude that my proposals can enormously help to diminish or solve the ethical problems of engineers and of technology. I also conclude that the implementation of my proposals, though feasible, requires important changes in several areas of the contemporary legal systems. These areas include liability law in general, corporate law, labor contract law, laws relating to secrecy and the duty to inform, and the procedures for collective decision making. (The latter issue has remained somewhat implicit in the present exposition.) The implementation of such changes would lead to more consensus about technological activities, among both actors and those subjected to the consequences. It would probably also lead to the decline of hierarchical organizations. Finally, this implementation would probably lead to a more sustainable development, and to better control of technological and other risks that threaten the continued existence of contemporary society.

ACKNOWLEDGMENT

I am indebted to prof. N. A. Ashford for critically reading a draft of these comments, especially Section 3 on developments in modern tort law.

NOTES

1. For what is said here on self determination, subjectivity, consent and liability, I am heavily indebted to Van Velsen (2000).

2. My present contribution focuses mainly on the implementation of strict(er) liability (iii). While doing that, I will have ample opportunity to illustrate (i). I will say a little on hierarchical organizations (ii), and nothing on collective decision making (iv). Neither will I say anything here on another essential topic, namely, that proper education of both engineers and the general public is necessary. At present, this education is in various respects inadequate or insufficient.

3. In some countries, e.g. the U.S.A., a distinction is sometimes made between real damages and so called punitive damages. Ashford (private communication, December 1999) has pointed out to me that in American law the distinction between liability for real damages and punitive damages is seen as follows. Real damages are related to costs of medical treatment and rehabilitation, property damages, and sometimes an award for pain and suffering. Punitive damages are meant as a punishment. According to Ashford, punitive damages are usually unrelated to actual damages and much larger in size. They are given for gross civil or criminal negligence. According to Ashford, punitive damages in American environmental and safety law are actually very important from the viewpoint of deterrence, and hence prevention. I will not go into this subject below.

4. See e.g. the relevant lemmas in the *Oxford Dictionary of Law* (Martin, 1994).

5. I do not imply here that at present, as a matter of actual positive law, tort law is relevant in all cases mentioned, but rather that tort law can/should be made relevant.

6. One source describing such developments in some detail is Kröner (1993), which provides an overview of developments and state of the art in environmental liability law in both North America and Europe. Another source, especially for European and Dutch developments, is Van Dunné (1993).

7. But as is pointed out later in the present comments, an agreement on the nature of liability for damage may of course be part of the conditions under which persons are willing to give their informed consent.

8. This evidence of financial responsibility is so to speak part of the actor's "license to operate", to use a term sometimes applied by industrial spokespersons in connection with their corporations' codes of conduct.

9. Elsewhere the author quoted in the text (Force, 1993) explains that in practice it will be very difficult for the responsible party to comply with the provisions that would limit liability under this Act. If this is true, the amount of financial responsibility actually required will *not* usually suffice to cover the actual liability. It hence remains possible that the responsible parties, though liable, nevertheless will fail to accomplish complete recovery. For instance, if the responsible parties are incorporated, shareholders are shielded from liability beyond the nominal value of their shares.

10. The italics have been added, because the phrase raises serious questions such as when expenditures are disproportionate, and how one may establish which expenditures are or are not disproportionate.

11. Compare Wicksell (1896) for a type of "contingent valuation methodology" applied to raising and spending tax money. Compare Van Velsen (1996) for a generalization of Wicksell's method.

12. For instance, the Dutch Civil Code treats micro organisms on a par with other environmental "toxics". See Messer (1994).

13. With respect to the subject of "small individual damage but large number of victims", a look at legal transformations that took place in the first quarter of the 19th century in the U.S.A. is very informative. See Horwitz (1977) for a detailed and very illuminating account. At least two aspects of these transformations are of particular interest here. One concerns the legal treatment of so called "consequential damages", being (usually small) injuries that were indirectly caused by industrial or economic activities such as watermills, roads, rail- and waterways. Another one concerns the development of the law of nuisance.

14. In the case of environmental damage, some questions may even be easier as compared to the two drivers case. For instance the question how to determine the share of each in the total liability: see the discussion of market share and pollution share liability above.

15. If, by contrast, complete repair were always possible, and still assuming the principle of self-determination, it might be maintained that strict liability *would* solve all ethical problems of engineers, if sufficient "evidence of financial responsibility" (see Section 3 above) would have been provided to repair the largest possible damage.

16. The Kyoto-protocol concerning the reduction of CO_2 exhaust, which extends to 2010, does not impose restrictions upon the CO_2 exhaust of developing countries. But if the climatic problem is to be brought under control, after 2010 such restrictions will certainly be necessary. See e.g. KNMI (1999).

17. I do *not* imply that these developments are not based upon, or conflicting with, "traditional ethics".

18. See *Verslag Parlementaire enquêtecommissie vliegramp Bijlmermeer* (1999). See especially the minutes of the interrogation of the El Al technician Gaalman and related interrogations.

19. For more on these historical views on allocating responsibility in organizations, see Bovens (1998, ch. 6).

20. As far as government organizations are concerned, serious flaws came to the fore during the Second World War. The lessons from Nazi Germany have not led to significant adjustment of the legal principles upon which hierarchical organizations operate. It should be noted, though, that the experiences with Nazi Germany were a major drive behind the introduction of the article on social responsibility in the Dutch law on higher education.

21. The transition from unlimited liability to limited liability of corporations took place roughly in the second half of the 19th century. See the various contributions in Orhnial (1982), and also Horwitz (1992, ch. 3).

22. This conditional liability was likewise in large part a product of 19th century. See e.g. Horwitz (1977).

23. A similar idea underlies the historical legal form known as the commenda or the societas maris (see Perrott 1982). This legal form developed from the partnership, where

the one partner, the commendator, provided the capital in the form of money or goods, and the other partner, the tractator, did or supervised the actual work (e.g., merchandising goods with a ship). The liability of the commendator was limited to his investment, whereas the tractator, whose share in the profit was 25%, had unlimited liability. Having its origin in Italy five centuries or more ago, the commenda lives on more or less unchanged in its modern guises known as societe en commandite, Kommanditgesellschaft, or commanditaire vennootschap. According to Perrott (1982), the commenda must be regarded as one of the most successful and popular forms of business association ever devised. But the last 100 or so years its importance has diminished greatly, because of the rise of the corporation with limited liability.

24. Davis (1998) has in essence suggested that engineers should have a veto right in certain matters (and not in others). In his opinion, engineers should, e.g., refuse to take orders that are inconsistent with "acting as a professional" (but should accept all other orders). This leads him to the question which factors (such as quality) are the professional's responsibility, and for which the engineer's judgement should therefore prevail. Unable to answer this question, Davis calls for studies that would help define these factors. It seems to me, however, that the non-hierarchical organizations working with "shares of responsibility" as depicted in the text do not require such explicit definitions for their proper operation. See also the review of Davis' book by D. Christiansen (1999).

25. Compare the following recent decision of an United States judge. According to the decision, the operator of the Internet site Blue Oval News was allowed to disclose information from internal Ford documents, from which it appeared that Ford sold fancy sports cars with defective engines, and that an upcoming Ford engine was unlikely to meet government emission standards. See URL: http://www.salon.com/tech/log/1999/11/10/copyright: "Fair use" vs. foul play.

REFERENCES

Bovens, M. A. P. (1998). *The Quest for Responsibility: Accountability and Citizenship in Complex Organizations*. New York: Cambridge University Press. (*Verantwoordelijkheid en organisatie. Beschouwingen over aansprakelijkheid, institutioneel burgerschap en ambtelijke ongehoorzaamheid*. Zwolle: W. E. J. Tjeenk Willink.)

Christiansen, D. (1999). Book review of Davis, Michael (1998). *Technology and Culture, 40* (October), 881–883.

Council for Science and Society. (1977). *The acceptability of risks* (p. 37). England: Barry Rose, Ringwood, Hants.

Davis, M. (1998). *Thinking Like an Engineer: Studies in the Ethics of a Profession*. New York:Oxford University Press.

KNMI. (1999). *De toestand van het klimaat in Nederland 1999*, redactie: G. P. Können. De Bilt: KNMI.

Force, R. (1993). Insurance and liability for pollution in the United States. Kröner (Ed.), (pp. 21–46).

Harrigan, M. J. (1993). Some concepts of collective responsibility in U.S. products liability and environmental situations. Kröner (Ed.), (pp. 47–50).

Horwitz, M. J. (1977). *The Transformation of American law, 1780–1860*. Cambridge, MA: Harvard University Press.

Horwitz, M. J. (1992). *The Transformation of American law, 1870–1960*. New York: Oxford University Press.

Kröner, R. P. (Ed.) (1993). *Transnational Environmental Liability and Insurance*. London: Graham and Trotman.

Messer, E. A. (1994). *Risico-aansprakelijkheid voor milieu-verontreiniging in het BW*. Thesis, Rijksuniversiteit Utrecht. Arnhem: Gouda Quint.

Martin, E. A. (Ed.) (1994). *A dictionary of Law* (3rd ed.). Oxford: Oxford University Press.

Orhnial, T. (Ed.) (1982). *Limited Liability and the Corporation*. London/Canberra: Croon Helm.

Perrott, D. L. (1982). Changes in attitude to limited liability – the European experience. Orhnial (Ed.), (1998), ch. 5.

de la Rue, C. (1993). Environmental damage assessment. Kröner, (Ed.), (pp. 67–78).

Sellschopp, H. D. (1993). *Multiple tort feasors/combined polluter theories, causality and assumption of proof/statistical proof technical insurance aspects*. Kröner, ed. (1993), 51–57.

Snijder. (1990). Van market share liability naar pollution share liability? *TMA/Environmental liability Law Review* (Rotterdam), quoted in Sellschopp (1993, ch. 6, 141–150).

Thompson, D. F. (1980). Moral responsibility of public officials: the problem of many hands. *APSR*, 74, 904–916.

Van Dunné, J. M. (1993). *Verbintenissenrecht, deel 2*, 2e herziene druk. Deventer: Kluwer

Van Velsen, J. F. C. (1996). Collectieve besluitvorming en vooruitgang. (Collective decision-making and progress). In: H. Zandvoort, (Ed.), *Technologie en verantwoordelijkheid (Technology and responsibility)*, reader gg 25 (pp. 121–126). Delft University of Technology.

Van Velsen, J. F. C. (2000). Relativity, universality and peaceful coexistence. *Archiv für Rechts- und Sozialphilosophie, 86*(1), 88–108.

Vaughan, D. (1996). *The Challenger Launch Decision: Risky Technology, Culture, and Deviance at NASA*. Chicago: University of Chicago Press.

Verslag Parlementaire enquêtecommissie vliegramp Bijlmermeer, 's-Gravenhage, 22 April 1999.

Wicksell, K. (1896). Ein neues Prinzip der gerechten Besteuerung. *Finanz theoretische Untersuchungen*, Jena. (An English translation is included in Classics in the theory of Public Finance, R. A. Musgrave & A. T. Peacock (Eds). London: Macmillan, 1967.)

Wilson, W. (1925) [1887]. The study of administration, reprinted in College and State:Educational, Literary and Political Papers (1875–1913). (*1*, pp. 130–158). New York.

Zweigert, K., & H. Kötz. (1987). *An Introduction to Comparative Law*. (2nd revised ed.). Oxford: Clarendon Press.

TOWARD A JURIDICAL TURN FOR THE ETHICS OF TECHNOLOGY? FINAL REMARKS

Ad Vlot

1. RECENT DEVELOPMENTS

In February 1999, during the time period in between my first response and these final remarks, a potential health hazard of our new material was discovered. All the aluminum surfaces in the GLARE material are covered with a primer that contains strontium chromate, a very carcinogenic chemical substance. This primer is still a common paint and bond primer for the aerospace industry and is applied for the surfaces to assure the durability of the bonding of the sheets in GLARE. It has been used in our lab for more than fifteen years. Most of the thin aluminum sheets are at this moment primed by spraying during production of the laminate at the Fokker company. The news of the dangers of the chromate began and spread like a rumor and not all the details are fully clear to me yet. The Dutch aerospace laboratory NLR temporally stopped specimen production after measuring too high a concentration in the air after some machine shop operations. The problem with chromate in primers is not new, and the aerospace industry has been using these chemicals for decades. Another part of the bonding process for GLARE that is based on chromate – the chromate acid anodizing of the aluminum before priming – is also still the standard pretreatment in the European aerospace industry. There is so much experience with these qualified processes that it is difficult to replace

The Empirical Turn in the Philosophy of Technology, Volume 20, pages 245–250.

ISBN: 0-7623-0755-2

them, for example, anodizing with the more environmentally friendly phosphoric acid anodizing process. New materials and technologies have a long way to go in aerospace technology before they become mature and accepted, mainly because of the emphasis on safety and the corresponding high costs necessary for qualification.

The new aspect of the GLARE technology is that the bonding process is not applied as a means to assemble parts, but as a way to create a new, laminated material, a semi-finished product. Unlike conventional bonded structures, this laminate may undergo more machining and cutting operations after the bonding process. Because there are also much more coated aluminum interfaces with primer in the laminated material, possibly a lot of chromate containing dust can spread in the air of the machine shop during extensive drilling, milling, and grinding operations. This dust may accumulate in the workshop.

Our research group was under extreme time pressure to meet the material selection milestone of the A3XX by the end of 1999. During the last period a lot of progress was made, but there were still a lot of technical hurdles, and there were some worries. A bigger problem was the management, the organization of this project. This large and highly specialized project is difficult to control financially, and also communication between all different people and partners is far from easy. All kinds of technical questions arise and it is hard to judge priorities. The new questions constantly disturb the time schedule which is never met anyway.

The news about the chromate sounded alarming to me. I contacted people at NLR to hear their story. They were waiting for the results of a measurement of the concentration in the air of their machine shop. Chemical experts told me that it was a dangerous substance, but that it is not clear whether it is dangerous when it is bound in the epoxy primer particles and spread during machining in the air as we do in our specific case. Our university health and safety expert explained to me that there are special regulations for this chromate. Others claimed that it was an old, well-known problem already described in the literature and that drinking coffee or smoking a cigarette would be more dangerous. The technical and organizational problems that we face at the moment tended to be more pressing than a vague health and safety issue. Stopping the specimen production in our lab at this moment on the weak basis of rumors would certainly kill the project.

Because the information was vague and inconclusive, we decided to start a subproject to clarify the problem by doing a literature search, by gathering the experience of other labs and companies, and by performing careful measurements to analyze the air in our lab. But this would take time. Weighing the evidence and the importance of the project was not straightforward. I was willing to stop specimen production but there was not sufficient evidence at this moment to support it. NLR did more measurements and did not find significant concentrations of chromates in the air, but our specimen production is much bigger and we have done it for a much longer period of time (fifteen years).

I tried to be frank and open on this. But I also experienced that due to the usual distortion in spreading rumors, the news that came back to me showed many characteristics of a sensational newspaper article, and claimed a problem that was over-exaggerated. I discussed the problems within our group and with my superiors. I also carefully and frankly shared the information that I had with the most important people who were involved at that moment: the workers in our machine shop. Their response was quite laconic. They were already tired by all the – in their eyes – stupid health and safety regulations that were ordained by the government and the university management during recent years. Besides, they were working already for fifteen years in this way, and the material had become a regular part of their life.

We also started a new project to try to replace the primer with a more sustainable one. But qualification and acceptance of such a primer in the material in the aerospace industry is certainly a long-term solution that will take many years.

I felt my responsibility, but acted as a small element and specialist within this large project in which so much is going on. In some way I received the informed consent of the workers in the machine shop, but is this enough to prevent liability? The NLR measurements were done and analyzed, and initial results were positive and reassuring. Later, in July 1999, measurements in our lab confirmed that the concentration of chromate in the air of the lab during operations was much lower than current regulations require. But measuring techniques and allowable concentrations as well as experiments (on rats?) to determine allowable concentrations are still under discussion. For chromate *every* dose – even if very small – implies a certain risk, a possibility, and how can you quantify and compare this risk? Are the tests carried out in our lab certain, representative, and accurate enough? And if workers become ill and relate their illnesses to this processing of the laminate, how can we prove that it was not caused by something else? At this moment in an objective and rational sense everything is OK and taken care of, but is it enough to silence my conscience? Is the manufacturer of the primer liable, or perhaps the government that banned the chemical too late? I love the creative engineering work and the co-operation between people to reach our goal (our new material on that plane (but the constant time pressure may lead to unforeseen risks and consequences.

During the same period (April 1999) another ethical aerospace case – also mentioned by Zandvoort – was in the headlines of the Dutch media. The results of an inquiry by the Dutch parliament of what happened after the crash in

Bijlmermeer (a part of Amsterdam) of a Boeing 747 cargo aircraft of El Al were revealed. The disaster was caused by a broken fuse pin due to fatigue in the mounting of one of the engines. The engine was separated from the wing and consequently led to damage and separation of another engine, and this chain of events led to damaged flaps of the wing which are important to keep the aircraft in the air at the low speeds necessary for a safe landing of the aircraft. The damaged flaps on one wing caused instability of the aircraft during the emergency approach to Schiphol airport. The El Al Boeing crashed on an apartment building and killed 39 inhabitants and the three crew members on board. The accident happened in 1992 and during the six years after the crash the rumors about the suspicious unknown cargo of the Israeli airliner became so strong that it affected public health and made the inquiry necessary. A group of inhabitants of Amsterdam and personnel of the fire brigade and the medical services suffered from a variety of symptoms. These were linked to the El Al cargo, and to the uranium that was used by Boeing as a counterbalance in the aircraft, and which was burned and spread in the air during the fire. The inquiry committee of the parliament was able to trace the contents of the cargo – which appeared to be almost a "mission impossible", given the reluctance of El Al and the Israeli government – and finally concluded that the cargo did not contain very dangerous goods. The government was accused of taking the complaints of the inhabitants of Amsterdam not seriously enough, and for creating enormous uncertainty for several years for the people involved. According to medical experts, this created emotional uncertainty, which perhaps caused (part of) the health problems. Is somebody liable, and who is liable for the vague health problems? El Al, the Dutch government, Israel, Boeing, the ministers, the engineers who selected uranium, El Al management? Is the relation between the health problems and the crash proven? What are exactly the health problems?

2. RESPONSE TO ZANDVOORT

(1) These cases illustrate that not only on the micro scale of my own university project, but also on the macro scale of the responsibility of organizations in society, liability is an extremely difficult matter. I support the "polluters pay" principle implemented in a complex juridical framework as Zandvoort proposes. I agree that this effort in itself may help to stimulate ethical awareness among engineers and managers, and may give health and safety issues a higher priority in technical development, for example, in our project to develop a new aircraft material. However, further juridical research is necessary, as Zandvoort admits. I do not see

light at the end of the tunnel at this moment, and I am not so sure about the outcome. United States experience with a juridical system based on a more strict liability has led to juridicalization and exorbitant claims. It sure led to a reduction of risks, e.g., one of four babies in the U.S. are delivered by a caesarian section because it is safer (Postman, 1993). And engineers sometimes choose well-known, bad solutions to prevent uncertainties.

(2) This juridical research should in my opinion be combined with an analysis of the empirical structure of modern technology and design, in its diversity among numerous technological disciplines. Probably not only a juridical change, but also a change in the way we develop and use technology and educate engineers is necessary. The solely juridical viewpoint can stimulate a kind of "end-of-pipeline" or black box ethics.

(3) I am not optimistic about the possibility of creating a tremendous book-keeping system to deal with risks in our technological society. The diffuse responsibility of numerous actors will be reflected in an equally diffuse system of funds and guarantors. Moreover, it is highly questionable whether most of the damage can be valued and quantified. The damage to the buildings due to the Amsterdam crash described above could easily be quantified and was recoverable, but what about the health problems? One of the students of an ethics course given at our university compared this type of approach with a "linear programming problem", by which a value is given by individuals and groups to (potentially) damaged property and landscapes, and the overall costs can be optimized quantitatively and damage can be compensated. Zandvoort's approach clearly reflects the modernistic worldview. The technical construction of an overall and global mechanism to control the risks in our society seems unrealistic to me. I am not so sure that it will increase R&D expenditures as Zandvoort claims, but I am sure that we will create a lot of work for lawyers and politicians.

(4) We are surrounded by a complex network of risks, which was rightfully named a "risk society" by Ulrich Beck (1992). (Who will compensate my stress and anxiety in this society?) The tasks for a new ethics with respect to modern technology in this type of society have been identified by Hans Jonas (1984), e.g., with respect to the diffuse relationship between actions and effects, and the global effects of modern technology over very long periods of time. Responsibility and liability have become a fundamentally obscure issue. To stimulate engineers to take responsibility for their actions in society the existence of laws, judges, and prisons is vital, and I support Zandvoort in this. But in order to create an open atmosphere and a responsible culture of care and sustainability in our society, more than a legal system to enforce liability is necessary.

(5) Punishment does not show us how to do good. The prohibition of murder will not create care. Apart from flaws in the juridical system, like the weak protection of whistle blowers, from my perspective three other flaws are at least equally important: (i) The education of engineers should cover in an integral way all the cultural aspects and background of technological artifacts. (ii) Engineering education should stimulate imagination, creativity, and an emphasis on the synthesis of all the different aspects and disciplines in design. (iii) One of the most serious flaws in modern engineering design method is the lack of direct contact of individuals behind the drawing board, in the laboratory, in the conference room or at the CAD-computer, with the actual effects during manufacturing and use of the product by people. This missing link with reality seems to me very fundamental for modern technology.

REFERENCES

Beck, U. (1992). *Risk Society, Towards a New Modernity*. London: Sage Publications.

Jonas, H. (1984). *The Imperative of Responsibility; In Search of an Ethics for the Technological Age*. Chicago: University of Chicago Press.

Postman, N. (1993). *Technopoly: The Surrender of Culture to Technology*. New York: Vintage Books.

NOTES ON CONTRIBUTORS

Davis Baird is Associate Professor and Chair of the Department of Philosophy at the University of South Carolina. He is the editor of *Techné: Journal of the Society for Philosophy and Technology*. He has published articles on a variety of topics in the history and philosophy of science and technology. He recently completed a book manuscript, *Thing Knowledge: A Philosophy of Scientific Instruments*, that should be published during 2001.

Bram Bos is a doctoral student at the Faculty of Philosophy of the Free University, Amsterdam. He is currently finishing his dissertation on control in technological systems, in which he tries to make a fruitful connection between philosophy of technology, on the one hand, and science and technology studies, on the other.

Louis L. Bucciarelli is a Professor of Engineering and Technology Studies at MIT. He is the author of *Designing Engineers* (1994), an ethnographic study of engineers at work, and, with Nancy Dworsky and Sophie Germain, *An Essay in the History of the Theory of Elasticity* (1980).

Philip Brey is Associate Professor in the Department of Philosophy of the University of Twente in the Netherlands. He received his Ph.D. from the University of California, San Diego. His research focuses on philosophical aspects of information and communication technology, including computer ethics, philosophy of artificial intelligence, and philosophical aspects of computer systems design. Recent publications are on worker autonomy and digital networks, the ethics of virtual reality, interdisciplinary method in computer ethics, design and the ontology of cyberspace and the implications of new media for the quality of life.

Clive L. Dym is the Fletcher Jones Professor of Engineering Design and chair of the Department of Engineering at Harvey Mudd College. He also serves as Director of the Center for Design Education. He has written ten books, including

Engineering Design: A Synthesis of Views (1994) and *Engineering Design: A Project-Based Introduction* (1999).

Peter Kroes studied physical engineering at the University of Technology Eindhoven (The Netherlands) and wrote his doctoral dissertation on philosophical problems concerning the notion of time in modern physical theories (University of Nijmegen, The Netherlands, 1982). Since 1995 he is Professor in Philosophy, specializing in the philosophy of technology, at the University of Technology Delft. Important publications: *Time: Its Structure and Role in Physical Theories* (1985), *Filosofie van de natuurkunde* (1987), *Technological Development and Science in the Industrial Age* (1992), and *Ideaalbeelden van wetenschap: Een inleiding tot de wetenschapsfilosofie* (1996).

Anthonie Meijers is Professor in the philosophy of technology at the Universities of Technology at Delft and Eindhoven. He studied mechanical engineering and philosophy and wrote his Ph.D. thesis on the foundation of the theory of speech acts in the theory of intentionality. Recently he edited *Belief, Cognition and the Will* (1999) and *Explaining Beliefs: Lynne Rudder Baker and her Critics* (forthcoming 2001). He is associate editor of the journal *Philosophical Explorations*. Together with Peter Kroes he is project leader of the research program "The Dual Nature of Technical Artifacts", carried out at Delft University of Technology and sponsored by the Netherlands Organization of Scientific Research.

Carl Mitcham is Professor of Liberal Arts and International Studies at the Colorado School of Mines. His recent books include *Thinking through Technology: The Path Between Engineering and Philosophy* (1994) and (with R. Shannon Duval) *Engineer's Toolkit: Engineering Ethics* (2002). He has also co-edited (with Stephen H. Cutcliffe) *Visions of STS: Counterpoints in Science, Technology, and Society Studies* (forthcoming 2001).

Joseph C. Pitt is Professor of Philosophy and Director of Graduate Studies in the Department of Philosophy at Virginia Tech. He has recently published *Thinking about Technology: Foundations of the Philosophy of Technology* (2000). Other books include *Galileo, Human Knowledge, and the Book of Nature* (1993) and *Pictures, Images, and Conceptual Change* (1998). He has edited or co-edited twelve other books and is the editor of the interdisciplinary journal, *Perspectives on Science: Historical, Philosophical, Social* published by MIT Press.

Arie Rip is Professor of Philosophy of Science and Technology at the University of Twente. He has published widely in the sociology of science and science policy, in technology dynamics and technology assessment. See for example, Rip, Thomas Misa, and Johna Schot, *Managing Technology in Society: The Paradigm of Constructive Technology Assessment* (1995).

René von Schomberg (Rene.von.schomberg@cec.eu.int) works as a scientific adviser for the European Commission on Socio-economic Issues of Science and Technology. He is author of consultative reports for United Kingdom's House of Lords Committee on EC regulation, the European Parliament, the European Environment Agency, and the Dutch Society for Bioethics. Recent edited books published in the English language include *Science, Politics and Morality, Scientific Uncertainty and Decision Making* (1993), *Contested Technology* (1995), *The Social Management of Genetic Engineering* (co-edited with Peter Wheale, 1998), *Democratising Technology* (1999), and *Discourse and Democracy, Contributions to Habermas "Between Facts and Norms"* (co-edited with Kenneth Baynes, forthcoming 2001).

Adrian Vlot is an Associate Professor and Deputy Chair of Aircraft Materials at the Faculty of Aerospace Engineering at Delft University of Technology, the Netherlands. He completed his Ph.D. dissertation entitled "Low-velocity Impact Loading of Fiber Reinforced Aluminium Laminates and Other Aircraft Materials" at Delft in 1991. He received his MA in philosophy from the Free University of Amsterdam in 1992. He is currently active in research regarding the impact damage phenomenon, fatigue, residual strength, repair and sustainable development of aluminium alloys and Fibre Metal Laminates. He gives lectures in Ethics, History of Technology and Philosophy. He was project leader of the GLARE Technology Development Project for the Delft University. One of his recent publications was on the Dutch philosopher van Riessen as philosopher of technology.

Henk Zandvoort is associate professor for ethics and technology at Delft University of Technology. He did his first degree in chemistry and in philosophy. His dissertation *Models of scientific development and the case of NMR* appeared in 1986 (Reidel, Dordrecht). Between 1986 and 1997 his main assignment was at the Dutch Ministry of Education, Culture and Sciences, as a civil servant in the area of higher education and research.

AUTHOR INDEX

Printed and bound by CPI Group (UK) Ltd, Croydon, CR0 4YY
16/09/2026
14963401-0003